Panzer 38(t)/Swiss LTL-H

Hans-Heiri Stapfer

Squadron Signal Publications

Covers and profiles by Don Greer
Illustrations by Matheu Spraggins

Introduction

The Panzerkampfwagen 38 (tschechisch), Armored Combat Vehicle 38 (Czech), abbreviated PzKpfw 38(t), was one of the most important Wehrmacht (German Army) tanks during the first half of World War II (WWII). The vehicle served in the Polish and French campaigns and in the invasion of the Soviet Union in the summer of 1941. The Soviets' T-34 battle tank quickly rendered the PzKpfw 38(t) obsolete, forcing the Wehrmacht to withdraw it from frontline combat in early 1942. The Germans continued to use the vehicle for anti-partisan operations, however, and there were still 229 PzKpfw 38(t)s in active Wehrmacht service in September 1944. A few even faced the Western Allies advancing into the Third Reich.

The PzKpfw 38(t) originated in the interwar years, when the Českomoravská-Kolben-Daněk (ČKD) firm in Praha-Libeň (a suburb of Prague) developed the Lehký Tank vzor 38 (light tank model 38 or LT vz 38) for the Czechoslovak Army, which ordered 150 vehicles on 22 July 1938. ČKD had produced only a handful when the Third Reich occupied and split up the Czechoslovak Republic in 1939. Germany realized the value of the Czechoslovak light tank and ordered an initial batch of 150 (chassis numbers 0001 through 0150) to be built between May and November 1939 at the Böhmisch-Mährische Maschinenfabriken AG (Bohemian-Moravian Machines Factory Limited or BMM), as the former ČKD was renamed after the German take over on 22 May 1939. Designated PzKpfw 38(t) Ausf A (Ausf is the abbreviation for Ausführung/model), 59 of the vehicles first saw action with the 67. Panzer Abteilung (67th Armored Battalion) of the 3. Leichte Division (3rd Light Division) during the Polish campaign.

BMM produced 110 PzKpfw 38(t) Ausf Bs (chassis numbers 0151 through 0260) between January and May 1940, followed by 110 PzKpfw 38(t) Ausf Cs (chassis numbers 0261 through 0370) between May and August 1940. In the Western campaign, 229 vehicles served with the 7. and 8. Panzer Divisions. BMM built 105 PzKpfw 38(t) Ausf Ds (chassis numbers 0371 through 0475) between September and November 1940. Based on lessons learned in the French campaign, BMM increased the front armor to 50mm and the hull and turret sides to 20mm. Between November 1940 and May 1941, BMM built 275 PzKpfw 38(t) Ausf Es (chassis number 0476 through 0750). The 250 PzKpfw 38(t) Ausf Fs (chassis numbers 0751 through 1000), built between May and October 1941, were nearly identical to the Ausf E. The PzKpfw 38(t) Ausf E/Fs served with earlier models in Operation Barbarossa. Seven Panzer Divisions (7th, 8th, 12th, 19th, 20th, 22nd, and 27th) fought against the Soviet Union. The PzKpfw 38(t) Ausf G was the last variant BMM built for the Wehrmacht. BMM produced 306 Ausf Gs (chassis numbers 1101 through 1359 and 1480 through 1526) between October 1941 and June 1942, with the unit price of a single late-production vehicle quoted at 56,000 Reichsmark.

Most PzKpfw 38(t) vehicles went to Germany, but some also saw action with the Reich's Eastern European satellites. Romania received 50 Ausf A through Ausf C vehicles. Bulgaria received 10 PzKpfw 38(t)s, designated *Praga*, which were used until 1952. In 1942 and 1943, the Magyar Királý Honvédség (Royal Hungarian Army) received 108 PzKpfw 38(t)s, which were redesignated T-38s. Most vehicles delivered to Hungary were late-production PzKpfw 38(t) Ausf E/Fs and Gs. Slovakia received a derivative of the PzKpfw 38(t), called LT 38. BMM initially built 37 vehicles for the Slovaks, and then Slovakia received an additional 37 vehicles from 1943 to 1944.

Prior to the German occupation in 1939, ČKD sold the Swiss Army 24 of the PzKpfw 38(t)'s predecessor, the Panzerwagen 39. The first batch arrived with gasoline engines, while the last 12 vehicles came in parts for Swiss assembly. These tanks were powered by Swiss Saurer CT-1D diesel engines and armed with a Swiss-designed 24mm gun and two 7.5mm machine guns.

About the Walk Around®/On Deck Series®

The Walk Around®/On Deck® series is about the details of specific military equipment using color and black-and-white archival photographs and photographs of in-service, preserved, and restored equipment. *Walk Around* titles are devoted to aircraft and military vehicles, while *On Deck* titles are devoted to warships. They are picture books of 80 pages, focusing on operational equipment, not one-off or experimental subjects.

ISBN 978-0-89747-589-1

Military/Combat Photographs and Snapshots

If you have any photos of aircraft, armor, soldiers, or ships of any nation, particularly wartime snapshots, please share them with us and help make Squadron/Signal's books all the more interesting and complete in the future. Any photograph sent to us will be copied and returned. Electronic images are preferred. The donor will be fully credited for any photos used. Please send them to:

Squadron/Signal Publications
1115 Crowley Drive
Carrollton, TX 75006-1312 U.S.A.
www.SquadronSignalPublications.com

Acknowledgments

Thanks to Dr. Attila Bonhardt, Ruedi Bühler, Bundesarchiv (Swiss Federal Archives) at Berne (Switzerland), Deutsches Panzer Museum at Munster, Walter Grube, Kerstin Gutbrod, Adj Uof Martin Haudenschild, Christian Hug, Lehrverband Panzer und Artillerie of the Swiss Armed Forces, Panzer Museum at Thun (Switzerland), George Punka, Wilfried Rorig, and the Tourist Office of Munster (Germany).

(Title Page) This PzKpfw 38(t) Ausf E/F, one of few surviving PzKpfw 38(t) vehicles, is on exhibit at the Deutsches Panzermuseum (German Armor Museum) in Munster, Germany. Between November 1940 and October 1941, 525 Ausf E/F vehicles were manufactured.

(Front Cover) This PzKpfw 38(t) Ausf C took part in the invasion of the Soviet Union in summer 1941. A number of bins and an unditching beam were mounted on the fenders. A Nazi flag was draped for aerial recognition.

(Back Cover) This Swiss Panzerwagen 39 (M-7542/hull number 10) belonged to the Panzerwagen Detachment 4. Its crew watches a German armored column passing close to the frontier between the Third Reich and Switzerland

A formation of Hungarian PzKpfw 38(t) Ausf Gs, belonging to the 30th Tank Regiment, are on the march to the Eastern Front. The closest vehicle is equipped with a rod aerial. A typical feature for late-production variants of the PzKpfw 38(t) was the location of the exhaust muffler, which was moved to a higher position on the rear armor plate. No bins were carried on the rear fender of this vehicle. This PzKpfw 38(t) carries Hungarian markings with a white and red outlined green cross. The Hungarian Army received a total of 108 PzKpfw 38(t)s from Wehrmacht (German Army) stocks during 1942. The Hungarian designation for the type was T-38. Hungarian crews were trained at the Military School at Wünsdorf in Germany. The Hungarian T-38 suffered very heavy losses against the superior Soviet T-34 tanks, and Hungarians withdrew the last PzKpfw 38(t) from frontline service in February 1943. (George Punka)

The PzKpfw 38(t) Ausf E/F is on exhibit next to a PzKpfw III Ausf M "Marlene" in the Deutsches Panzermuseum at Munster. The PzKpfw 38(t) is painted in Dunkelgrau RAL 7021 (dark gray) overall. Dunkelgrau RAL 7021 became the official vehicle color scheme of the Wehrmacht in late 1939, and all PzKpfw 38(t)s built after this date left the paint shop in the BMM factory, at Praha-Libeň, in RAL 7021. The PzKpfw III Ausf M on the right is painted in Gelbbraun RAL 8000 (yellow brown). RAL 8000 was applied for the first time during March 1941 on vehicles scheduled for delivery to the Deutsches Afrika Korps (German Africa Corps). North Africa was the sole theater of operation where the PzKpfw 38(t) did not serve, and thus, it was never painted Gelbbraun RAL 8000. During operation on the Ostfront (Eastern Front), a number of PzKpfw 38(t)s received a washable white winter camouflage paint or had, as a substitute, a rough coat of lime (calcimine) whitewash applied. Since the whitewash supplies were often limited, many PzKpfw 38(t)s only had parts of the vehicle painted or used patterns of stripes or spots in white. Dunkelgelb RAL 7028 (dark yellow) became the new base color for German vehicles during 1943, but PzKpfw 38(t) production ended in June 1942, so PzKpfw 38(t)s never left the factory painted Dunkelgelb. Some vehicles were, however, repainted in Dunkelgelb RAL 7028 over the original Dunkelgrau RAL 7021. Other PzKpfw 38(t)s had irregular patches of Dunkelgelb RAL 7028 applied over the original camouflage.

This frontal view of the PzKpfw 38(t) Ausf E/F appears against the backdrop of the museum's collection of helmets from various countries. On the hull in front of the crew compartment on the PzKpfw 38(t) Ausf E/F was a straight front plate that provided an increased armor protection of 50mm and featured an enlarged left visor. The PzKpfw 38(t) Ausf E/F was the first version with a component to accommodate carrying a large section of spare track on the lower front armor plate. The PzKpfw 38(t) Ausf E/F was also the first variant equipped with two equal-sized visors. Earlier versions had a smaller visor for the radio operator, who sat on the left side. The apertures for the MG 37(t) 7.92mm machine gun in the front left hull plate and the front right plate of the turret were faired over by circular cover plates. Most PzKpfw 38(t)s carried an MG 37(t) machine gun at both these locations. A white Y has been painted on the left edge of the front plate on this vehicle. This Y is the divisional marking of the 7. Panzer Division that saw combat during the French campaign in May and June 1940 as well on the Ostfront. The sword-wielding ghost figure seen to the right of the vehicle's right visor is the unofficial emblem of the 11. Panzer Division (known as the Gespensterdvision or "Ghost Division"), a unit that never had the PzKpfw 38(t) in strength. The vehicle's nickname, *Wilde Sau*, (wild sow) appears above the left visor. The practice of naming individual vehicles was not as widespread in the Wehrmacht as it was in the American and British forces.

PzKpfw 38(t) Design Evolution

PzKpfw 38(t) Ausf A

The PzKpfw 38(t) Ausf A adopted the original Czechoslovak communication aerial, which was fitted in an armored tube that ran along the left side of the hull. This antenna was removed during operation. The Ausf A was not supplied with a Notek light on the production line, but a number of vehicles were modified with a Notek blackout light on the left front fender and a Notek convoy light atop the left rear engine decking. The early Ausf A did not have a splash ring around the MG 37(t) 7.92mm machine gun, which was mounted in the turret. A solid cap was wrapped around the suspension's semi-elliptical leaf springs.

PzKpfw 38(t) Ausf B

The Czechoslovak communication aerial tube was eliminated from all PzKpfw 38(t) Ausf Bs at the production line, but the vehicles were fitted with a Notek blackout light on the left front fender and a Notek convoy light on the left rear engine decking. All Ausf Bs still carried the original Czechoslovak whip aerial, and most vehicles had a splash ring around the MG 37(t) 7.92mm machine gun, which was mounted on the right side of the turret. A handle was fixed to the left side of the turret. The semi-elliptical leaf springs' cap had an oval aperture.

PzKpfw 38(t) Ausf C

The PzKpfw 38(t) Ausf C's exhaust muffler was moved to a higher position on the rear armor plate. The Ausf C was the first variant equipped with the German rod aerial, which replaced the Czechoslovak whip aerial. A bump stop was fitted to the front of the first road wheel. The rear Notek convoy light moved from the upper left decking to the top of the left rear fender. Late-production Ausf Cs had the rear-view mirror eliminated from the left front fender.

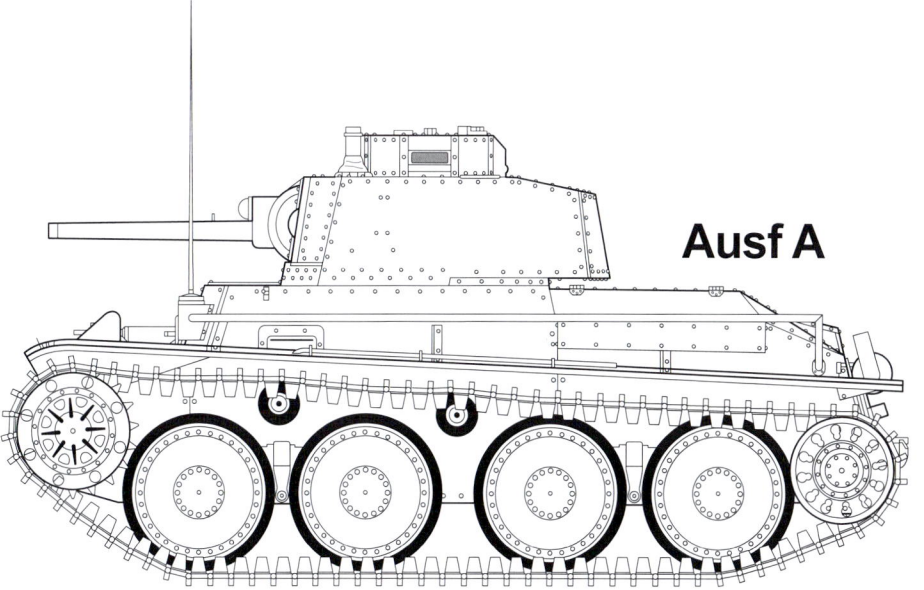

Ausf A

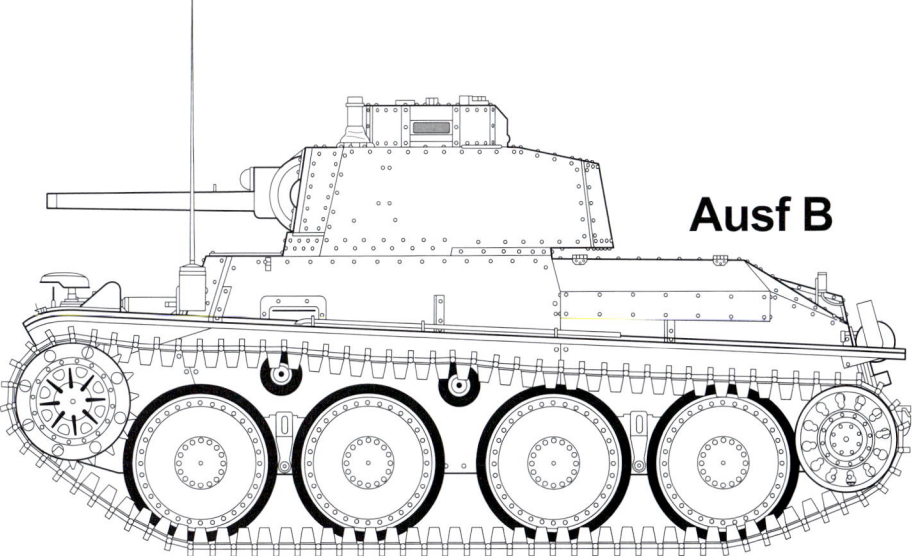

Ausf B

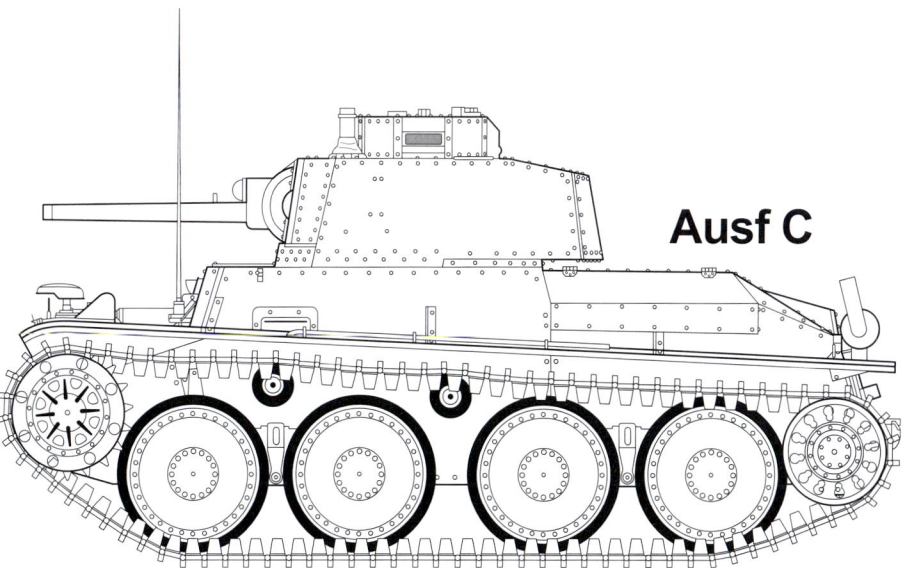

Ausf C

PzKpfw 38(t) Ausf E/F

The left-side visor on the front plate was enlarged to match the right-side visor on the PzKpfw 38(t) Ausf E/F. A straight front plate was fixed on the hull to give the crew compartment a total of 50mm of frontal armor protection. The vehicle could carry a single large section of spare track on the lower front armor plate and two small spare tracks on each side of the upper armor plate, in front of the crew compartment. A gun sight was mounted in front of the driver. The rear-view mirror on the left front fender and the splash ring around the turret-mounted MG 37(t) were eliminated at the production line, but a smoke generator box was mounted on the left rear armor plate. Each spring for the road wheels had 15 leaves—one more than previous PzKpfw 38(t) versions.

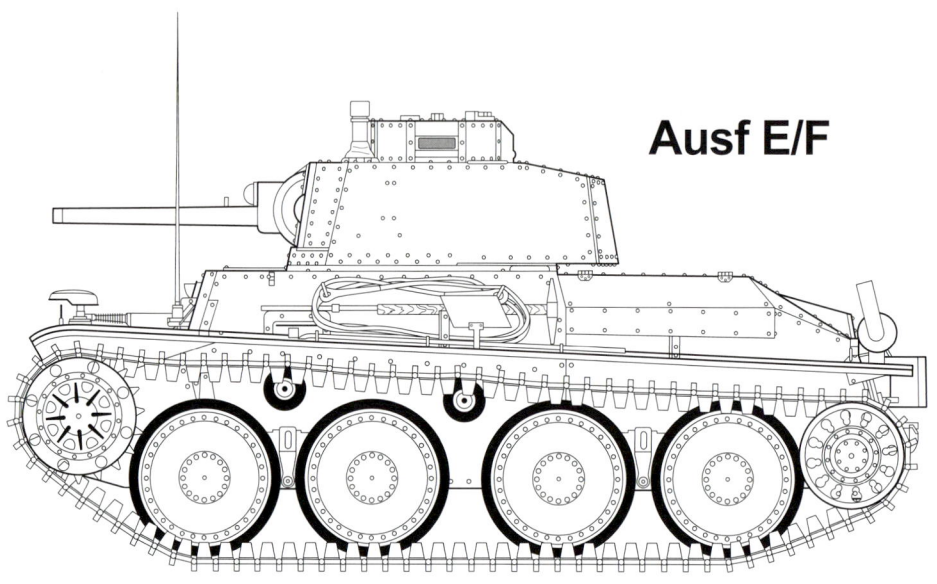

Ausf E/F

Ausf G

PzKpfw 38(t) Ausf G

The Notek blackout light moved to the upper left, front edge of the hull on the PzKpfw 38(t) Ausf G. Most Ausf Gs had a reinforcing cap sans oval aperture that wrapped around the semi-elliptical leaf springs. A circular, red reflector was mounted atop both rear fenders, and a stop light also was added to the right rear fender. The two track tension adjusters, located on the rear armor, received a semicircular cover. The number of rivets on the engine bay access panels was reduced. The external starting crank on the rear armor plate received a conical covering. The engine air exhaust received a slide to regulate the airflow.

A large access panel was incorporated into the armor plate on the upper hull front of the PzKpfw 38(t). This Ausf E/F has an upper plate thickness of 30mm. Earlier models only had a 12mm upper armor plate. This access panel allowed access to the differential box that contained the two drive sprocket clutches and brakes. The differential box was part of the Praga-Wilson Type DV-TNHP pre-selector gearbox.

The PzKpfw 38(t)'s access panel remained unchanged throughout the vehicle's production cycle. Two hinges are on the upper edge of the panel.

A U-shaped splash ring surrounds the access panel to the Praga-Wilson Type DV-TNHP pre-selector gearbox. All vehicle versions received this splash ring. The PzKpfw 38(t)'s hull was riveted. The rivets are in front of the splash ring. Early PzKpfw 38(t)s had 12mm-thick armor on the access panel, but the Ausf E/F and G models had 30mm-thick armor. The splash ring was introduced for the first time on the LTP tank for Peru.

This hinge is located on the top left of the access panel. Further to the left of the photograph is the upper edge of the U-shaped splash ring.

The PzKpfw 38(t)'s right front fender was the same design as the left fender. Both fenders remained unchanged throughout the entire production cycle of the vehicle. Some vehicles were equipped with a circular white reflector on top of each front fender, but this PzKpfw 38(t), on exhibit at the Deutsches Panzermuseum at Munster, lacks these reflectors.

An L-bracket connects the lower front armor plate with the covering for the drive sprocket.

All PzKpfw 38(t)s had tow hooks riveted toward the right and left ends of the lower front armor plate.

The right drive sprocket of the PzKpfw 38(t) is seen here. Steering involved the use of two horizontally mounted levers in the driver's compartment. Pulling either lever disengaged the appropriate clutch and activated the brake to turn in the selected direction.

This PzKpfw 38(t) Ausf E/F's upper fender, on the left side is unusual for a typical Ausf E/F variant because this example lacks the Notek blackout light that was attached to the top of the left front fender. With the exception of the Ausf A and Ausf G, all other PzKpfw 38(t) models carried the Notek blackout light on the left front fender.

Reinforcement brackets, like this one on the left front fender, helped reduce fender damage. Rough front-line conditions frequently left the sheet metal fenders bent upward from impacts with obstacles. The brackets on the left and right fenders mirrored each other and remained unchanged throughout the production life of the PzKpfw 38(t).

The vehicle's left track is engaged with the front drive sprocket. The track links and their headless track pins consisted of manganese-nickel steel. Each side of the vehicle had 94 track links.

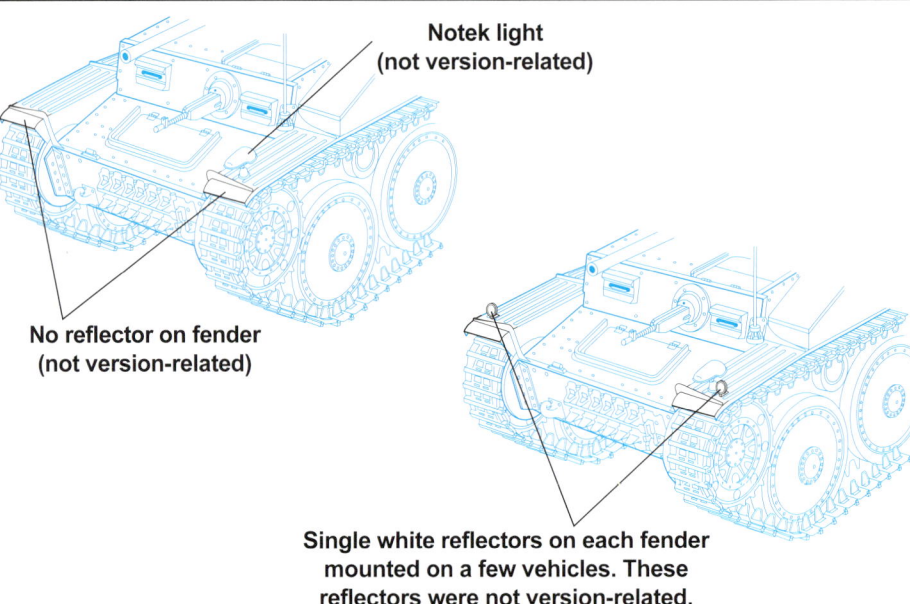

Notek light
(not version-related)

No reflector on fender
(not version-related)

Single white reflectors on each fender
mounted on a few vehicles. These
reflectors were not version-related.

The PzKpfw 38(t) Ausf E/F had a component that allowed the vehicle to carry a seven-link section of spare track on the lower front 50mm armor plate. Earlier PzKpfw 38(t) models did not have this feature.

A boom secured the spare track on the lower armor plate. The support arm was not directly mounted on the armor plate. Instead, the support arm was welded on a reinforcement plate that was attached to the 50mm armor plate. The early vehicles did not carry spare track and did not have these two enforcement plates.

The left-side reinforcement plate connected the spare track holder's support arm to the lower 50mm armor plate. Rivets attached the hook to the lower armor plate. On the Swiss LTL-H design, the hook was located in a slightly lower position on the lower front armor plate.

The tow hook's retention plate, on the lower, right side of the tank, extends and attaches to the lower, back-slanted armor plate. This small, rectangular armor plate has a thickness of 50mm on the Ausf E/F and all subsequent versions of the PzKpfw 38(t). The lower horizontal armor plate that covers the chassis of the tank has a thickness of 8mm. The lower chassis armor remained 8mm thick through the production cycle of the PzKpfw 38(t).

The left front section of the PzKpfw 38(t) has circular armor protection for the inner covering of the drive sprocket. The left front section lacks the four rivets that were found on the right side of the vehicle next to the access panel.

The vehicle's lower-left front has a riveted, back-slanting lower armor plate that is 50mm thick. Most armor plates were manufactured by the Komotauer Edelstahlhütte AG, the former Poldihütte or Poldina Hut' (Poldi Ironworks) at Komotau (now Kladno, Czech Republic). Six large and two small rivets attach the tow hook to the hull.

A ram-shaped tow hook that points to the right is fixed to the right front section of the PzKpfw 38(t). This assembly was a feature on late-production PzKpfw 38(t)s. The four rivets just between the fender and the front access panel are found only on the right side of the vehicle. A gunsight was mounted on many PzKpfw 38(t)s in place of the four rivets. When the gunsight was removed, the four rivets filled the holes.

An L-shaped bracket connected the lower armor plate with the left-side circular armor protector of the sprocket wheel covering. This PzKpfw 38(t) had just three rivets on the face that connected the bracket to the lower armor plate. Most late-production models were mounted with five rivets. The inner circular cover plate has been fixed in place by three bolts plus a rivet mounted on top and another on the bottom of the rear of the L-shaped bracket. Four rivets on the side facing the inner circular cover plate replaced these two rivets and three bolts on late-production models.

This hook attaches to the lower left front armor plate. The hooks were used for recovery purposes and rated at 5,000 kilograms. Recovering vehicles and tanks from muddy roads was a frequent task for Panzer Divisions assigned to the Eastern Front.

The PzKpfw 38(t)'s cast manganese-steel track shoes were connected by pins held in place by spring clips that allowed rapid changing.

Ausf A

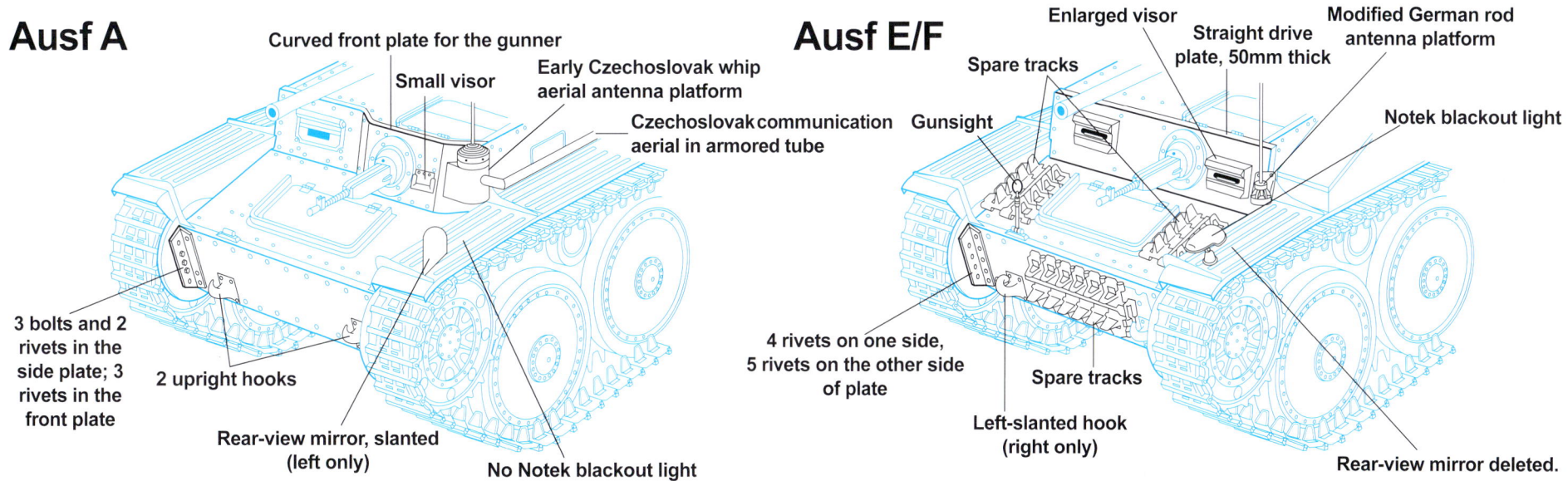

Curved front plate for the gunner

Small visor

Early Czechoslovak whip aerial antenna platform

Czechoslovak communication aerial in armored tube

3 bolts and 2 rivets in the side plate; 3 rivets in the front plate

2 upright hooks

Rear-view mirror, slanted (left only)

No Notek blackout light

Ausf E/F

Enlarged visor

Straight drive plate, 50mm thick

Modified German rod antenna platform

Spare tracks

Gunsight

Notek blackout light

4 rivets on one side, 5 rivets on the other side of plate

Spare tracks

Left-slanted hook (right only)

Rear-view mirror deleted.

The rear-view mirror housing was attached to the front fender with four screws. The circular mirror was protected by a covering located on the rear housing.

The rear-view mirror housing of a Swiss LTL-H is shown here in the closed (left) and open position (right). Touching a button on top of the housing opened the cover.

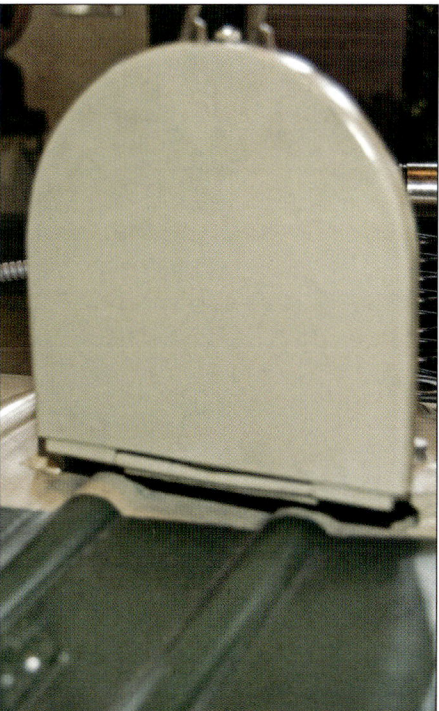

PzKpfw 38(t) Ausf A

Back

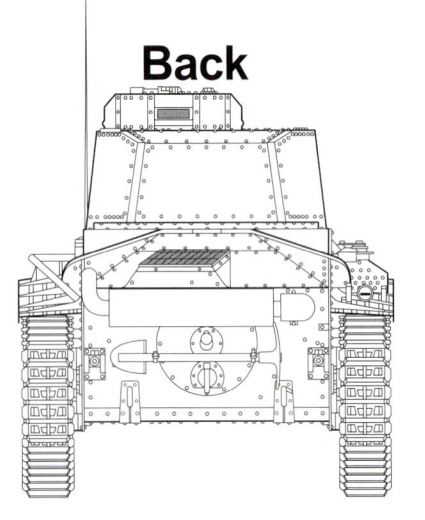

Top

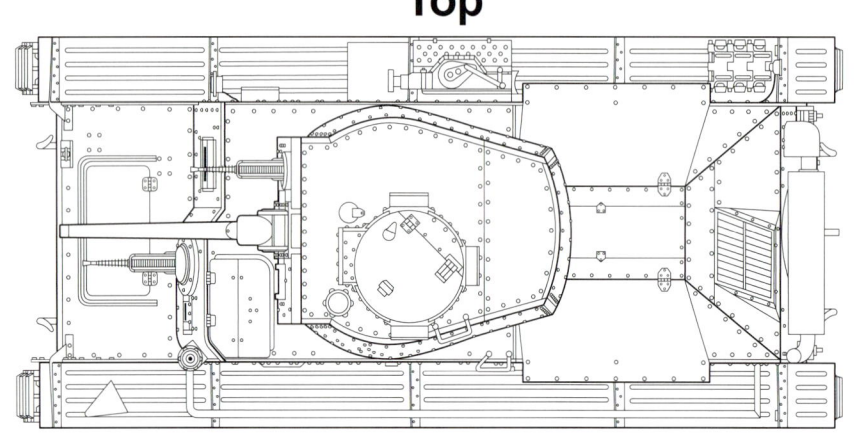

Front

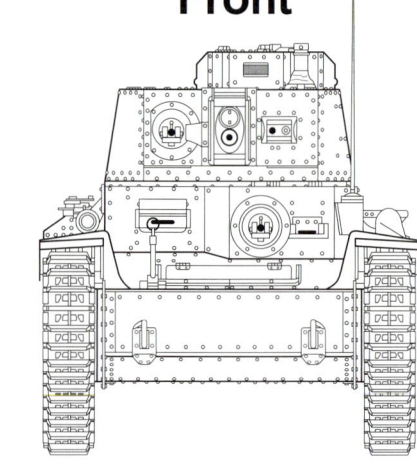

Length	4,610 meters
Width	2,135 meters
Height	2,252 meters
Weight (loaded)	9,725kg
Powerplant	Praga TNHPS/II six-cylinder inline gasoline 7.754cc with 125hp at 2,200rpm
Transmission	Praga Wilson pre-selector gearbox with five forward and one reverse gears
Speed	42km/h on road, 17km/h off road
Range	250km on road, 100km off road
Armament	1X Škoda Kwk 28(t) 37.2mm gun (90 rounds) 2XMG 37(t) 7.92mm (2,700 rounds)
Crew	3 to 4

Left

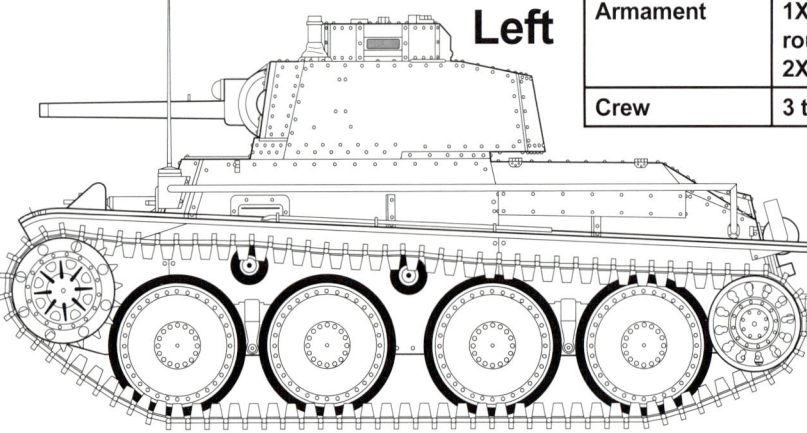

Right

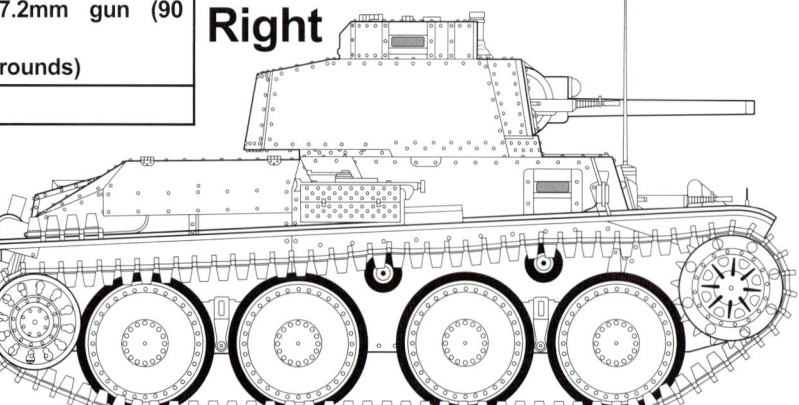

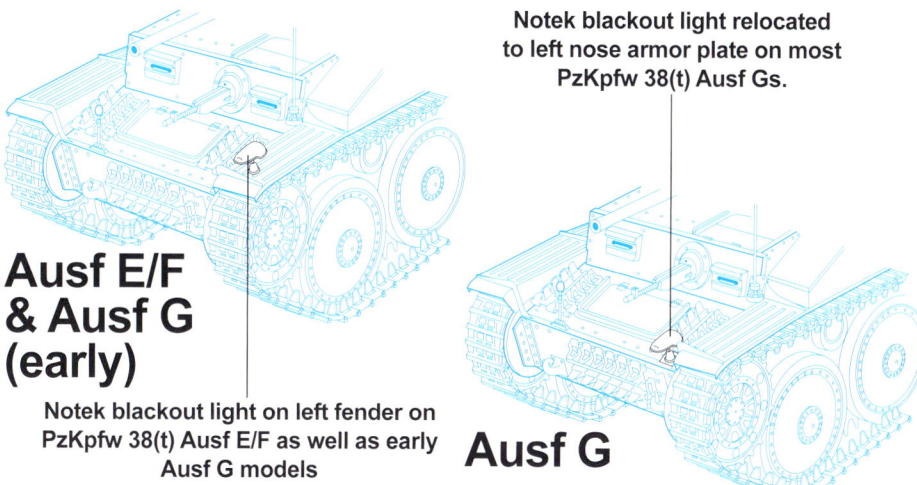

Notek Light Development

Notek blackout light relocated to left nose armor plate on most PzKpfw 38(t) Ausf Gs.

Ausf E/F & Ausf G (early)

Notek blackout light on left fender on PzKpfw 38(t) Ausf E/F as well as early Ausf G models

Ausf G

The disc-shaped armored covering for the sprocket wheel is fixed to the front fender on both the left and right sides of all PzKpfw 38(t) versions. Two rivets attached the reinforcement frame to the upper front armor plate.

The right fender reinforcement is wedge-shaped. The Swiss LTL-H, Peruvian LTP, and pre-war Czechoslovak LT vz 38 had longer, curved fender extensions.

The inside of the left fender of this PzKpfw 38(t) lacks the Notek blackout light that was usually mounted on all wartime PzKpfw 38(t)s. The aperture for the MG 37(t) 7.92mm machine gun has been covered by a circular piece of armor. With the exception of the commando vehicles, all PzKpfw 38(t) were equipped with the MG 37(t) machine gun. This weapon could be fired by the driver using a Bowden cable from one of the steering levers. Part of the U-shaped splash ring in front of the access panel is also visible.

15

One of the big advances of the Czechoslovak-designed PzKpfw 38(t) over contemporary German designs was its advanced suspension. It consisted of four large rubber-tired single wheels on each side. Each wheel was mounted on a stub axle with semi-elliptical springs for each pair of wheels. The rubber-tired single wheels had a 77.5cm diameter. A 6mm-thick armored disk covered each wheel, making the wheels themselves an incoming fire protection for the hull, the springs, and the swing arms. These wheels had the same effect as the side Schürzen (skirts) that the Wehrmacht mounted on the PzKpfw III and PzKpfw IV later in the war. In contrast to the TNH (Persia), the LTL-H (Switzerland), and the LTP (Peru), the crew compartment was stretched further forward on the PzKpfw 38(t). The vehicle featured visors on both sides of the vehicle for the driver (right side) and the radio operator (left side). Although it bears insignia of the 7. Panzer Division that operated against the Red Army, this particular vehicle lacks the stowage bins that were very common on PzKpfw 38(t)s deployed to the Ostfront. Two armored panels cover the PzKpfw 38(t)'s engine bay, which was located behind the crew compartment and contained the Praga TNHPS/II powerplant. This particular PzKpfw 38(t) also lacks the horn that was attached behind the right side visor on all versions of this light tank. This vehicle had been delivered to Sweden during WWII and was presented by the Swedish government to the German Panzer Museum.

(Top Left) The front drive sprocket had 19 teeth that transported the manganese-nickel steel track with headless track pins. The design of the front drive sprocket remained the same through the production cycle of the PzKpfw 38(t). Eight circular apertures were in the outer sprocket as well eight arch-shaped apertures in the inner sprocket.

(Top Right) Eight bolts attached a circular cover plate onto the front drive sprocket. The cover plate could be removed for maintenance or replacement of the entire sprocket. Located in the center is a nipple for lubricating the bearings with grease.

(Bottom Right) The arch-shaped aperture in the inner front drive sprocket was introduced on the PzKpfw 38(t). The previous LTL-H and LTP tank designs for Switzerland and Peru, from which the PzKpfw 38(t) developed, had 12 circular apertures in the outer sprocket.

The front drive sprocket teeth engaged the slots in the track lines and assured proper transmission of engine power to the tracks. Each track link had two guide horns that ensured the tracks would remain centered on the road wheels.

This front drive sprocket is on the vehicle's right side.

This inner front drive sprocket is on the right side of the vehicle.

The 19 teeth in the front drive sprocket engaged slots in the track links in order to propel the track, which was 29.3cm wide. The PzKpfw 38(t) attained a speed of 48 km/h on the road.

(Above) Two horns on each link of the track ensured that the road wheels remained centered on the track. The return rollers measured 22cm in diameter and were fitted with rubber tires. A leaf spring served two road wheels.

(Top Left) Longitudinal ridges were stamped into the fenders to enhance the sturdiness of the metal. The front fenders remained unchanged through the PzKpfw 38(t)'s production cycle.

(Bottom Left) The left front fender has an L-bracket riveted to it.

The right-side drive sprocket with the first pair of road wheels and the return roller are seen here on this PzKpfw 38(t). Short cranks were used to mount the wheels in pairs. A semi-elliptical leaf spring served each pair of wheels and was fixed to the hull on a single central mount. A bump stop with an inverted V profile is mounted on the hull just in front of the first road wheel on both this side and the left side of the vehicle. This bump stop was introduced for the first time on the PzKpfw 38(t) Ausf C and remained throughout the rest of production cycle of the PzKpfw 38(t). The early PzKpfw 38(t) Ausf A and Ausf B models had no bump stop.

This manganese nickel steel track is what the PzKpfw 38(t) used.

The road wheels with solid rubber tires had a diameter of 77.5cm. The rim was fixed to the road wheel by 32 rivets. This type of road wheel and the suspension arrangement were subsequently adopted on the Jagdpanzer 38(t) Hetzer in a slightly modified version. The road wheels on the Jagdpanzer 38(t) Hetzer were 10cm larger in diameter, but the number of rivets on the outer rim was reduced to 16 on the late Hetzer variant.

The circular panel was fixed to the wheel with 16 bolts. This covering could be removed for maintenance purposes. A nipple in the center of the road wheel was used to lubricate the bearings with grease. The road wheel remained unchanged through the production cycle of the PzKpfw 38(t).

The semi-elliptical leaf springs on this Swiss LTL-H (Panzerwagen 39) have the same configuration as the early-production German PzKpfw 38(t) models. Ausf A through Ausf C versions had only 14 leaves in the leaf springs, and the ends of the springs were not cropped. Ausf B through Ausf E/F versions had an oval aperture on the face of the cap that wrapped around the semi-elliptical leaf springs.

All the springs mounted on the late-production variants of the PzKpfw 38(t) Ausf E through G had 15 leaves in order to counterbalance the additional weight caused by additional armor protection. Most PzKpfw 38(t) Ausf G series vehicles had a reinforcing cap wrapped around the semi-elliptical leaf spring without the oval aperture that was standard on the PzKpfw 38(t) Ausf B trough Ausf F.

Semi-elliptical leaf springs with 15 leaves had a solid covering and cropped spring ends.

The bump stop was attached on the lower hull section in front of the first road wheel.

These road wheels are mounted on the semi-elliptical leaf spring. Above the spring is the return roller. The spring's front face cover is solid, which is not standard for a PzKpfw 38(t) Ausf E/F, which usually has an oval aperture in the front face of the steel covering.

The two guide horns on each track link ensure the track stays centered on the return rollers. All German PzKpfw 38(t)s had two return rollers on each side. The German PzKpfw 38(t) had a thicker rubber tire fitted on the return rollers as compared to the rubber tire on the pre-war LTL-H and LTP types.

Two return rollers were attached on each side of the hull. The design of the return rollers was not changed throughout the production cycle of the PzKpfw 38(t). These return rollers all had "rubber" tires made from Buna, which was a synthetic rubber produced from brown coal.

These return rollers were attached by five bolts on the hull of the PzKpfw 38(t). The Jagdpanzer 38(t) Hetzer continued to use this type of return roller, but the tank hunter derivative of the PzKpfw 38(t) only had a single return roller on each side.

The return rollers on the PzKpfw 38(t) prevented the first pair of road wheels from being touched by the upper track.

The track drops over the right rear pair of road wheels because the tank does not have a third return roller. This design gave the PzKpfw 38(t) track a distinctive appearance as compared to the LTP and LTL-H tanks for Peru and Switzerland and the LT 40 for Slovakia, all of which had a third return roller attached on the rear hull.

The two rear road wheels lacked a return roller. The third return roller, included on the LTL-H and LTP tanks, which were predecessor designs for the PzKpfw 38(t), prevented the track from touching the top of the two rear road wheels.

The road wheel and idler configuration on both sides of the vehicle remained the same on all models of the PzKpfw 38(t). The idler had a diameter of 53.3cm. Each track of the PzKpfw 38(t) had a length of 9.77 meters.

The circular apertures on the left idler wheel of a PzKpfw 38(t) suggest that the idler wheel previously saw service with a SdKfz 138 Grille or an early production batch of the Jagdpanzer 38(t) Hetzer.

The circular apertures are evident in the outer idler wheel. A nipple is located in the center of the circular access panel that can be removed for inspection purposes.

The right idler of the PzKpfw 38(t), located towards the rear of the vehicle, had no teeth, unlike the drive sprockets and their 19-toothed rings located on the front of the vehicle. Circular apertures on the outer idler were not standard on any PzKpfw 38(t). Most vehicles had apertures consisting of two unequally sized holes merged together to form "figure 8s." A few PzKpfw 38(t)s were equipped with idler wheels that featured arch-shaped apertures. The kind of idler on this particular PzKpfw 38(t) Ausf E/F was not mounted on the BMM production line, but it subsequently became standard on the SdKfz 138 Grille (cricket), an artillery close-fire support platform for mechanized infantry units, as well as on the early-production batches of the Jagdpanzer 38(t) Hetzer.

The idler wheel is photographed from below. The inner disc of the idler also had circular apertures. The lower rear armor plate had a thickness of 8mm. Like the entire hull, the lower part of the PzKpfw 38(t) was riveted.

The swing arm for the right idler wheel was connected with the track tension adjuster located on the rear armor plate of the PzKpfw 38(t). Swing arms for the idler wheels on both sides of the vehicle were identical on all versions of the PzKpfw 38(t).

These two rings are on the left idler wheel on the PzKpfw 38(t).

The 29.3cm-wide track wraps around the left idler wheel of the PzKpfw 38(t).

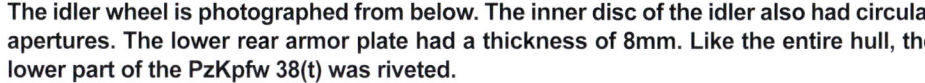

Idler Wheel Development

Arch-shaped aperture on some PzKpfw 38(t) Ausf C, D, and E/Fs.

Keyhole-shaped apertures, standard on all variants

Standard

Non-standard

25

The PzKpfw 38(t) was armed with a single KwK 38(t) 37.2mm cannon, developed by Škoda, with a 4,000-meter range and a muzzle velocity of 750 meters per second. KwK 38(t) gun rounds could penetrate armor as thick as 35mm. Ninety rounds for the KwK 38(t) were carried in 15 spring-loaded magazines, and 2,700 rounds were carried for the two air-cooled MG 37(t) 7.92mm machine guns. The Kampfwagenkanone (armored vehicle gun) (KwK) was as effective as the cannon installed on contemporary PzKpfw III versions. Experienced crews could fire at a rate of 15 rounds per minute. An MG 37(t) 7.92mm machine gun was mounted on the right side of the turret. The turret could traverse 360 degrees by way of a hand traversing gear, which could be disengaged and the turret moved by the gunner using the gun's shoulder pad. The turret housed the gunner, who was on the right side of the vehicle, and the commander, who was on the left. The MG 37(t) machine gun has been removed from this particular PzKpfw 38(t), and a circular cover plate is riveted over the aperture for the barrel. A visor was located on each side of the hull for the driver and the radio operator. Atop of the turret is a copula featuring four visors set at 90-degree angles from one another. Next to the copula on the turret is a periscope. Although operational PzKpfw 38(t)s had a horn located atop the fender just in front of the right visor, this particular vehicle lacks the horn. Neither the turret nor the crew compartment of the PzKpfw 38(t) was gas proof. The six white rings on the gun barrel represent the kill tally.

The PzKpfw 38(t) had a 50mm-thick flat, front armor plate to protect the driver and the radio operator. Earlier models had a staggered front plate with only 25mm of armor protection. The circular plate covers the station for the MG 37(t) 7.92mm machine gun, which became standard on operational PzKpfw 38(t)s.

The left cast visor obviously suffered some battle damage. The visors were only lowered during combat since they provided a rather limited field of view for the driver and the radio operator. On road marches, a windshield could be inserted, and it protected the driver and the radio operator from wind and dirt.

The left visor for the radio operator is on the front plate of the PzKpfw 38(t) Ausf E/F. On late-production models with the flat front armor plate, the left and right visors were identical in size. On the earlier PzKpfw 38(t) Ausf A through C, the left visor for the radio operator was smaller than the visor on the right.

Stepped oval slits in the visor of the PzKpfw 38(t) allowed limited vision. The stepped aperture deflected lead splash and bullets from the 4mm wide slit. A 50mm-thick block of bullet-proof armor glass could be inserted on the inside of the visor, which could also be fully opened to give the crew a much-improved field of view.

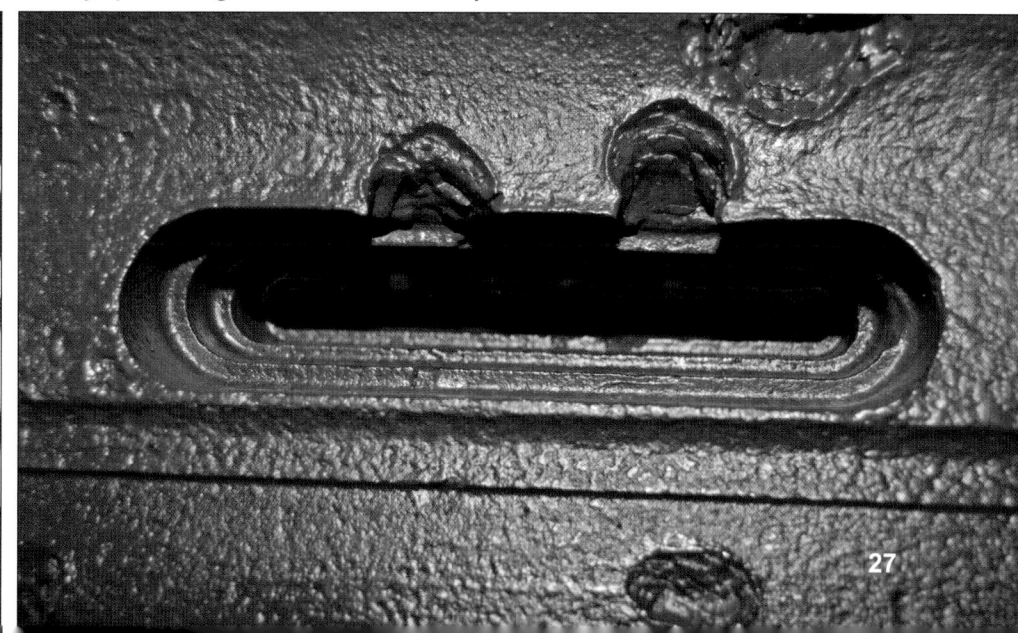

The right-side visor housing was larger than the left visor and was mounted at a higher position on the superstructure than was that on the left. This PzKpfw 38(t) does not carry the horn that was normally mounted on the L-shaped fender bracket. The massive 50mm-thick front hull plate that had been introduced on the PzKpfw 38(t) is clearly visible in this picture.

The large right-side visor housing for the driver was riveted to the superstructure of the PzKpfw 38(t). This type of visor was not changed throughout the production cycle of the PzKpfw 38(t). The LTL-H and LTP design for Switzerland and Peru lacked such a housing and instead had a horizontal slit incorporated in the hull.

The left visor for the radio operator is smaller than the right-side visor housing and is located in a lower position, just above the fender. The pre-war LTL-H and LTP designs lacked the left-side visor altogether. The longitudinal ribs in the sheet metal fender enhanced its strength. This particular PzKpfw 38(t), on exhibit at the Deutsches Panzermuseum, also lacks the German rod aerial that normally was attached on the front side armor plate.

The small left-side visor for the radio operator was much smaller than the visor on the vehicle's right side. This size difference was a feature of all versions of the PzKpfw 38(t). The left visor was located at a lower position than the right visor.

The exit hatch is located above the left-side of the crew compartment. The two-door hatch is connected with hinges.

The PzKpfw 38(t) had two exits—one above the left side of the crew compartment and the other on the copula of the turret.

Four L-brackets fixed each fender to the superstructure of the PzKpfw 38(t). The fenders are riveted to the L-brackets.

The non-riveted part of the L-bracket was always mounted to face the front. The fenders on all PzKpfw 38(t) were made from sheet metal.

A circular armor plate fares over the aperture for the 7.92mm MG 37(t) machine gun in the turret's right front plate. Most vehicles had an MG 37 rather than the plate.

The splash ring around the machine gun aperture was deleted from all late-production PzKpfw models as can be seen in this view of the right side of a late-production turret.

The 37.2mm KwK 38(t) cannon had an elevation of + 25 degrees and a depression of - 10 degrees. A hand wheel attached to the gun cradle was used for the manual elevation and depression of the cannon. The fairing above the barrel covers the recoil cylinder for the KwK 38(t).

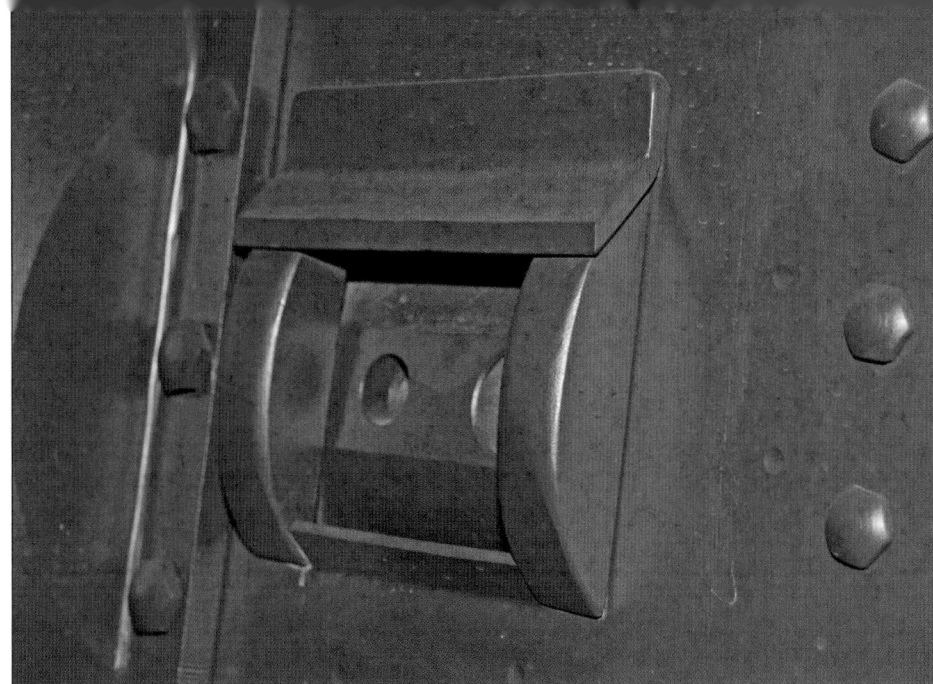

The two circular apertures in the Turmzielfernrohr 38(t) gun sight hold the sighting lenses for the tank commander. These two lenses, which have different diameters, follow the depression or the elevation movement of the 37.2mm KwK 38(t) cannon. This sight has a magnification of 2.6 times and a 25 degree field of vision.

A housing covers the sighting lenses and prevents them from being obstructed by rain and snow.

The commander occupied this position on the left side of the turret. Atop the turret is a copula with four episcopes mounted at 90 degree angles from each other. The coaxial Turmrundblickfernrohr 38(t) (periscope) for the commander rotates 360 degrees. The front armor plate of the turret was 50mm thick on late-production versions, such as on this PzKpfw 38(t) Ausf E/F. The PzKpfw 38(t) Ausf E/F also had fewer rivets on the front turret extensions than earlier versions.

31

The turret's side armor on late-production models was 15mm thick. Large vehicle numbers were common for tanks assigned to the 7. Panzer Division. The fully-rotating steel turret sat on a ball bearing race that was 126.5cm in diameter. A bin to hold a camouflage net, jack, and wooden jack base was mounted at the base of the vehicle's fender at the BMM factory in Praha-Libeň. The bin is missing on this vehicle.

In the front turret section, the commander aimed the gun through the coaxial telescopic sight with 2.8-times magnification and elevated it using either a shoulder control or an elevating gear. Each of these was locked automatically in position when the gun was fired. The front plates for the turret were set at an angle of 10 degrees.

The left side of the turret of the PzKpfw 38(t) on exhibit in the Deutsches Panzermuseum at Munster is seen here. The observation copula for the commander is not centerline mounted, but offset to the left. The side armor of the turret is sloped at an angle of 9 degrees from vertical. The previous LTL-H and LTP designs for Switzerland and Peru lacked such a copula with four episcopes. The vehicles instead had a visor on the right and left sides of the turret.

Splash Ring Development

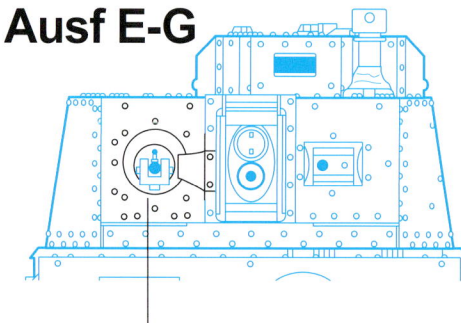

Ausf E-G

Splash Ring discontinued on the front turret armor plate

Ausf B-C

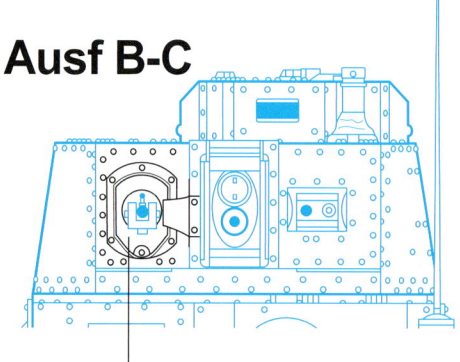

Splash ring mounted on right front turret in front of the MG 37(t) 7.92mm machine gun.

The commander's Turmrundblickfernrohr 38(t) (periscope) was mounted on the roof of the turret, in front of the observation copula next to the front episcope. In contrast to the remaining three episcopes on the two sides and in the rear, the face of the front episcope was entirely vertical. The remaining three episcope coverings were sloped toward the top.

The Turmrundblickfernrohr 38(t) (periscope) was bolted to the turret. Most all operational PzKpfw 38(t)s had a leather covering wrapped around the lower half of the periscope to keep dirt, rain, and snow from getting into the periscope's ball mounting. This particular vehicle does not have the leather cover.

The commander could rotate the coaxial Turmrundblickfernrohr 38(t) (periscope), mounted on the upper left of the turret, 360 degrees. The periscope had a magnification of 2.6 times and a 25-degree field of view. The armored cap had two rows of ball bearings allowing it to move freely with the periscope. The armored cap was unique to the German PzKpfw 38(t). The periscopes used on the pre-war LT vz 38 of the Czechoslovak Army, on which the PzKpfw 38(t) was based, lacked this armored cap and had only a thin periscope head.

This view of the rear of the observation copula reveals its left-of-center mounting atop the turret. This particular vehicle does not have the horizontal handle that was fixed on the upper edge of the sloped side armor. Except for the Ausf A, all other variants had this handle on the left side.

This episcope is one of four mounted in the copula. With the Turmrundblickfernrohr 38(t) (periscope), these episcopes provided the commander with all-around visibility. All episcope upper faces were sloped, except the front one, which was vertical.

The copula hatch was one of the two main exits from the PzKpfw 38(t). A single, circular hatch is fixed with two hinges on the copula. The roof of the turret as well as the hatch had an armor thickness of 15mm on the late-production versions of the PzKpfw 38(t). The circular covering in the decking of the turret can be opened to fire rockets or to wave signal flags.

The two hinges on the copula hatch were of different sizes from each other on all PzKpfw 38(t). A grasp had been added on the copula.

Standard

No rear stowage bin.

Stowage bin is non-standard,
not version-related.

Non-standard

The front episcope on the copula has a vertical, rather than sloped, face.

Compared with the LTL-H and LTP tank designs for Switzerland and Peru, the right rear of the turret on the PzKpfw 38(t) Ausf E/F had been extended, giving extra space for the 15 spring-loaded magazines for the KwK 38(t) gun and the ammunition boxes for the MG 37(t) machine gun in the turret bustle. The rear turret section extends over the crew compartment. The extended turret prevents the armored engine bay cover doors from opening unless the turret was turned 90 degrees to either side to allow access.

This view of the back of the turret shows clearly that the observation copula was located to the left of the center of the turret. This location reflected the position of the commander inside the vehicle on the left side of the turret. A 30mm-thick armor plate covered the back of the PzKpfw 38(t) turret. In contrast to the front episcope, the upper part of the covering of the rear episcope is sloped. The LTP design for Peru and the TNH for Persia had a visor located in the rear armor plate of the turret. Apart from these two types, no other version of the Czechoslovak light tank ever carried a visor in the rear armor plate of the turret.

Two Hungarian PzKpfw 38(t) Ausf E escort an PzKpfw IV. There is a single bin located on the right fender and spare track links are stored on the rear fender. The exhaust muffler has been moved to a higher position and a smoke generator box is mounted on the left rear armor plate. The Notek convoy light is mounted on the left fender beside the muffler. There is a tactical number on the rear of the turret, but not on the side armor plate. (George Punka)

Two MG 37(t)s are mounted in the turret and the front armor plate of this whitewashed PzKpfw 38(t) belonging to the Hungarian Army. The right visor and the left access hatch on top of the superstructure are open. A horn is mounted atop the right fender. A typical feature on late-production PzKpfw 38(t)s was the storage of spare track links on the left and right of the upper armor plate. A feature of the PzKpfw 38(t) Ausf G was the mounting of the Notek blackout light in front of the left spare track link. (George Punka)

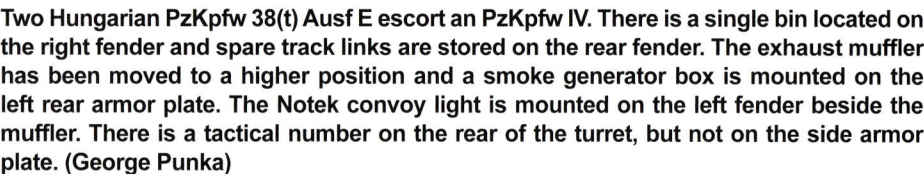

Bin Variations-All Versions

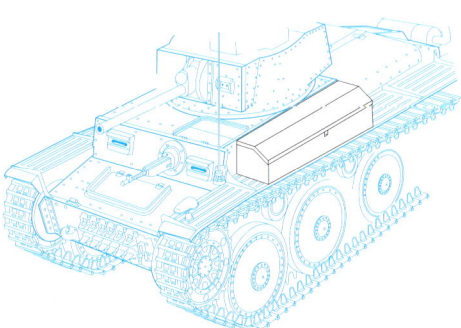

Single Large Bin

Two Bins

Three Bins, Different Sizes

Two Bins, Unditching Beam

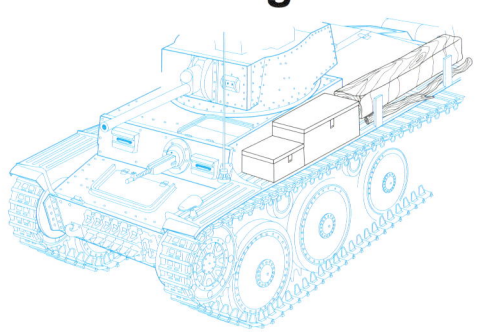

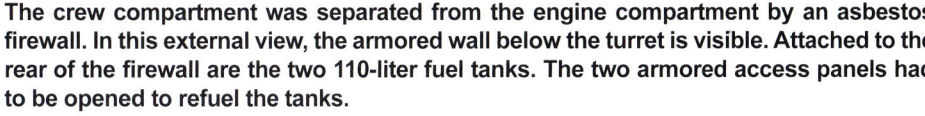

The crew compartment was separated from the engine compartment by an asbestos firewall. In this external view, the armored wall below the turret is visible. Attached to the rear of the firewall are the two 110-liter fuel tanks. The two armored access panels had to be opened to refuel the tanks.

The rear of the left, armored engine access panel shows the two horizontal rows of rivets that were a feature of all PzKpfw 38(t)s except the PzKpfw 38(t) Ausf G and late-production models of the PzKpfw 38(t) Ausf F. These two horizontal rows of rivets were deleted on all PzKpfw 38(t) Ausf Gs as well the late-production PzKpfw 38(t) Ausf Fs, so that only the vertical riveting on both edges of the access panel remained.

The engine compartment was covered by two large armored engine-bay panels. Air to cool the engine was drawn through openings in the armored engine access panel overhang. Two 110-liter fuel tanks located on each side of the engine compartment gave the vehicle a range of 220 kilometers. Passive fire protection was provided by special bolts holding the engine compartment floor underneath the gasoline tanks. These bolts were supposed to give in case of fuel explosion so that burning gasoline would shoot outside the vehicle and not into the crew compartment.

The outer firewall separates the crew compartment from the engine compartment. The armored engine-bay access panel is riveted as are the hull and the turret.

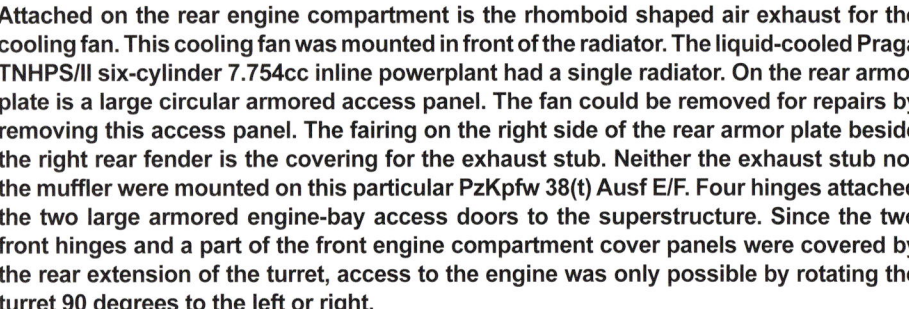

The armored engine bay access doors that were attached to the upper vehicle armor with two massive hinges for each individual door. All PzKpfw 38(t) had two hinges per door, while the LTL-H for Switzerland had three hinges per door. The doors had 8 mm armor protection.

The massive hinges that connected the superstructure and the armored access panels for the engine bay. There were a total of four hinges attached on the engine compartment. Only the two rear hinges were visible, since the forward hinges were covered by the extension of the turret.

Attached on the rear engine compartment is the rhomboid shaped air exhaust for the cooling fan. This cooling fan was mounted in front of the radiator. The liquid-cooled Praga TNHPS/II six-cylinder 7.754cc inline powerplant had a single radiator. On the rear armor plate is a large circular armored access panel. The fan could be removed for repairs by removing this access panel. The fairing on the right side of the rear armor plate beside the right rear fender is the covering for the exhaust stub. Neither the exhaust stub nor the muffler were mounted on this particular PzKpfw 38(t) Ausf E/F. Four hinges attached the two large armored engine-bay access doors to the superstructure. Since the two front hinges and a part of the front engine compartment cover panels were covered by the rear extension of the turret, access to the engine was only possible by rotating the turret 90 degrees to the left or right.

The upper engine compartment was riveted as were the hull and turret.

With the exception of the Ausf G, all PzKpfw 38(t)s were equipped with this rhomboid air exhaust on the upper rear of the engine deck. On the Ausf G, a slide was introduced on the left-side frame that could regulate the airflow passing through the tunnel from the radiator. Freezing conditions on the Ostfront made the innovation necessary. This kind of slide subsequently became standard on the Jagdpanzer 38(t) Hetzer tank hunter.

The rhomboid air exhaust tunnel was bolted to the rear engine deck. The grille kept out foreign debris that could cause damage to the fan or the radiator of the Praga TNHPS/II engine. The rhomboid-shaped air exhaust tunnel was a feature that was typical of the PzKpfw 38(t); the tunnel on the previous LTL-H design for Switzerland was square in shape.

Four reinforcement ribs were stamped into the sheet metal used on the rear fenders of all German PzKpfw 38(t)s. In contrast, ribs in an X pattern were stamped into the rear fender metal of the Swiss LTL-H tank. A total of four L-brackets on each side connected the fender to the hull of the PzKpfw 38(t). Clean rear fenders were a rarity during operations. Most vehicles had spare tracks or Jerry cans mounted on top of the fender. All German PzKpfw 38(t)s had straight ends, while the original Czechoslovak pre-war LT vz 38 had a rounded downward end plate on the rear fender.

Four longitudinal ribs reinforce the right fender on the PzKpfw 38(t).

This view is of the left rear fender of the PzKpfw 38(t) on exhibit in the Deutsches Panzermuseum at Munster. This particular vehicle lacks the Notek convoy light that was attached on the left rear fender on all operational PzKpfw 38(t)s. With the exception of the early PzKpfw 38(t) Ausf A and Ausf B, all models of this tank carried the Notek convoy light on the base of the rear fender.

The overhang of the armored engine compartment cover doors is evident. All versions of the German PzKpfw 38(t) had four longitudinal stabilization ribs stamped in the fender, while the Swiss LTL-H tanks had three X-shaped ribs stamped into the rear fender.

This otherwise very neatly restored PzKpfw 38(t) Ausf E/F on exhibit in the Deutsches Panzermuseum lacks parts that were essential on wartime vehicles. No muffler is mounted on this particular vehicle, and the Notek convoy light is missing. All vehicles belonging to the Wehrmacht had to be equipped with the Notek convoy light. The light should be located on top of the rear fender. This vehicle also lacks the smoke generator box mounted on all operational PzKpfw 38(t) Ausf E/F on the left upper rear armor plate. This particular vehicle also lacks the towing rope, crow bar, axe, and shovel that were mounted on the base of the fender in a position below the turret at the production line of the BMM factory. Almost all PzKpfw 38(t)s that saw action on the Ostfront (Eastern Front) carried a number of bins, Jerry cans, and other equipment on the fender. The Balkenkreuz (beam cross) national symbol would either be applied on the rear armor plate of the vehicle or on the rear armor plate of the turret. There were also a number of vehicles that never got any national or divisional markings. Divisional markings, such as the ghost insignia on the right of the rear armor plate, were seldom applied on German tanks.

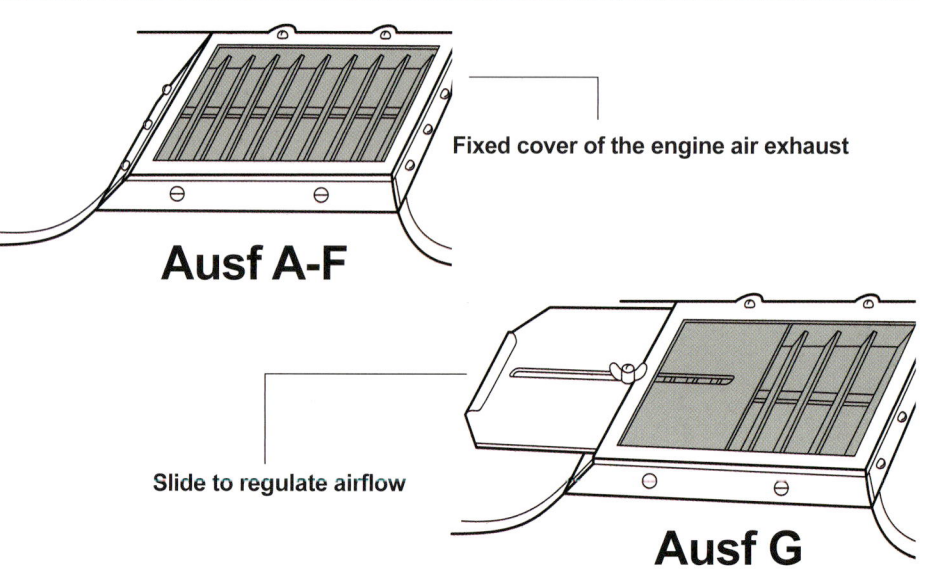

Fixed cover of the engine air exhaust

Ausf A-F

Slide to regulate airflow

Ausf G

The rhomboid-shaped air exhaust is on top at the rear of the engine compartment. Only late-production Ausf Gs had a slide that regulated the radiator's air flow. All other models, including this vehicle, lacked this slide.

This frame is for the air exhaust that services the radiator fan of the water-cooled Praga TNHPS/II engine. The frames on the left and right were different. The PzKpfw 38(t) Ausf A to Ausf F had this type of air exhaust with internally adjustable louvers.

The left side of the frame for the rhomboid-shaped air exhaust became typical for the PzKpfw 38(t) Ausf A to Ausf F.

PzKpfw 38(t) Ausf G

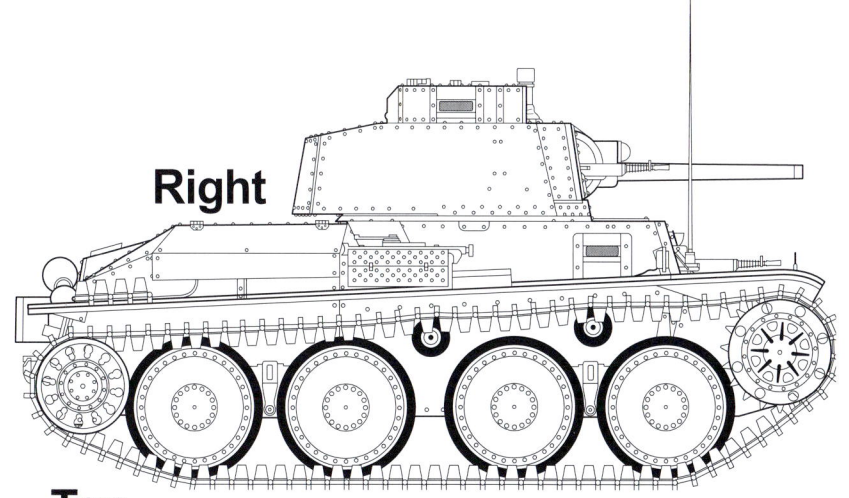

Right

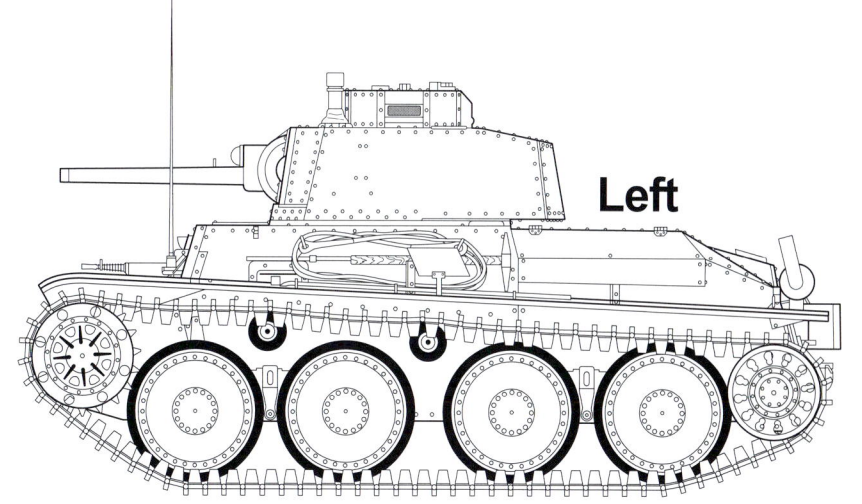

Left

Top

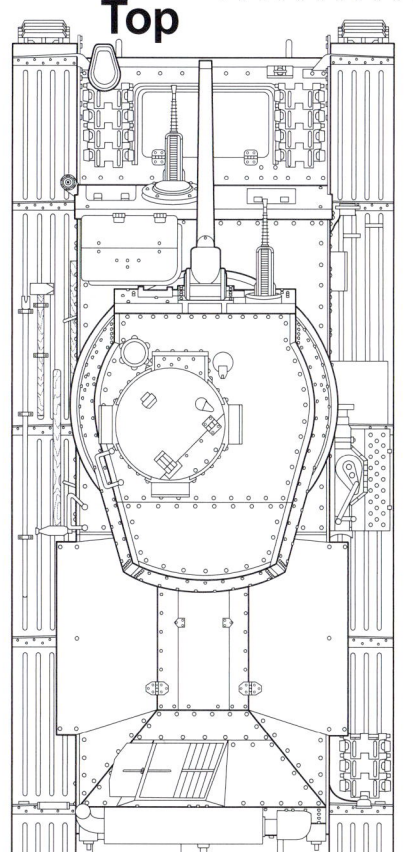

Length	4,610 meters
Width	2,130 meters
Height	2,252 meters
Weight (loaded)	10,354kg
Powerplant	Praga TNHPS/II six-cylinder gasoline, 7,754cc with 125hp at 2,200rpm
Transmission	Praga-Wilson pre-selector gearbox with five forward and one reverse gear
Speed	42km/h on road, 17km/h off road
Range	220km on road, 100km off road
Armament	1xKwk 38(t) 37.2mm gun (90 rounds) 2xMG 37(t) 7.92mm machine gun (2,700 rounds)
Crew	4

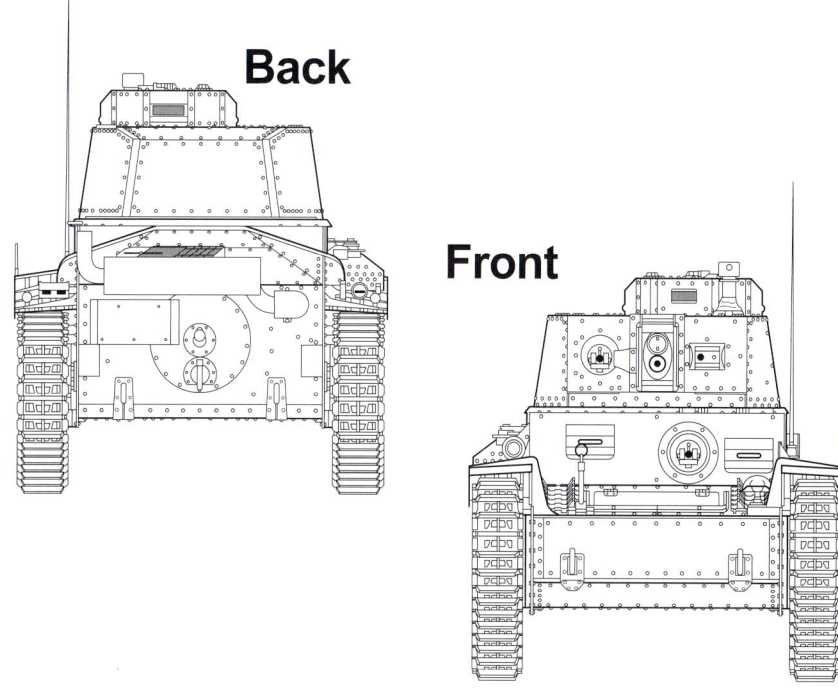

Back

Front

The rear armor plate of a PzKpfw 38(t) had a thickness of 15mm. The track tension adjusters were located on the left and right edges. The late-production model PzKpfw 38(t) Ausf G featured a semicircular covering for these track tension adjusters. Here, a circular armor plate covers the fan that cools the radiator of the inline Praga TNHPS/II gasoline engine that had an output of 125 HP at 2,200 revolutions per minute. Mechanics could gain quick access to the fan by removing the circular armor cover plate. The German PzKpfw 38(t) and the Czechoslovak LT vz 38 were the first vehicles to be equipped with a circular access hatch for the fan. Previous LTL-H and LTP models for Switzerland and Peru lacked such an access hatch. The Praga TNHPS/II engine was started electrically with a Bosch starter motor. If necessary, a hand crank could be used to start the powerplant. A corresponding shaft for the crank was located in the center of the circular armored panel. This crank ran through the fan and directly to the engine. A conical cover for the shaft became a distinctive feature of the PzKpfw 38(t) Ausf G, but the earlier PzKpfw 38(t) Ausf A through Ausf F versions had no such cover.

A triangular extension was mounted on top of both the left and right rear fenders. This extension remained unchanged throughout the production cycle of the PzKpfw 38(t).

The left rear fender with the four longitudinal reinforcing ribs stamped into the sheet metal is here seen from below.

The back end of the fender on all PzKpfw 38(t)s was straight. This design was in sharp contrast to the Czechoslovak pre-war LT vz 38 design that featured a downward-curved fender extension. The PzKpfw 38(t) Ausf A through Ausf F versions lacked the stop light on top of the right rear fender that became a standard feature on the PzKpfw 38(t) Ausf G. This final version of the PzKpfw 38(t) also had a red, circular rear reflector mounted on the right and left rear fenders. Most earlier variants of the vehicle lacked these reflectors, as does the vehicle on exhibit in the Deutsches Panzermuseum.

Muffler Development

Ausf A/B

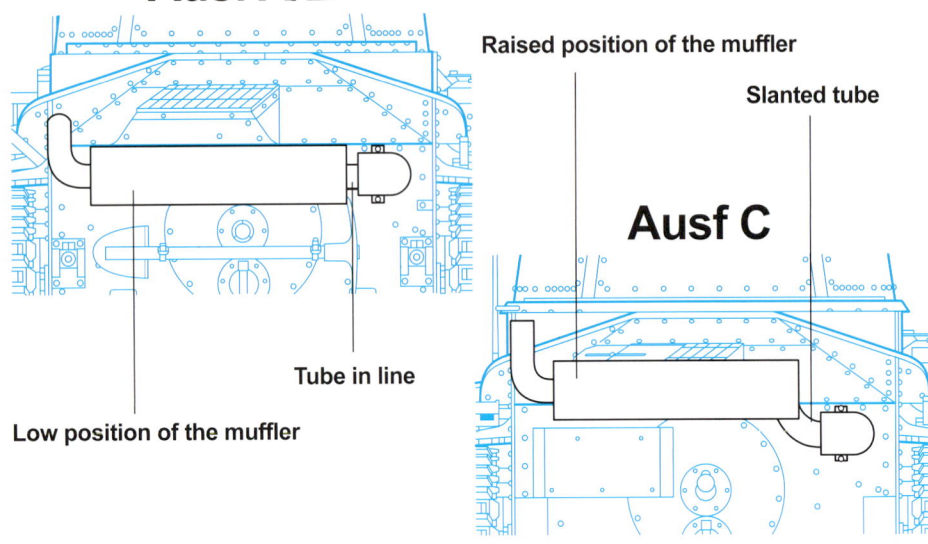

Raised position of the muffler

Slanted tube

Ausf C

Tube in line

Low position of the muffler

The metal covering for the muffler was located on the right rear armor plate of the PzKpfw 38(t). This type of covering is from a Swiss LTL-H (Panzerwagen 39), which was very similar to all German PzKpfw 38(t)s.

The muffler of the Swiss LTL-H (Panzerwagen 39) was similar to the early-production German PzKpfw 38(t) Ausf A and Ausf B variants. The muffler on these vehicles was located in a low position, but on the PzKpfw 38(t) Ausf C through G variants, the muffler was mounted at the top of the rear armor plate.

This object is the left track tension adjuster. With the help of a swing arm connected with the idler wheel, turning the screw either direction properly adjusted the track. Four bolts attached the track tension adjuster to the end plate.

Track tension adjusters were attached on the left and right sides of the outer rear end plate of the PzKpfw 38(t).

This object is the right track tension adjuster of a PzKpfw 38(t). These track tension adjusters remained the same on the PzKpfw 38(t) Ausf A through Ausf F variants. On the PzKpfw 38(t) Ausf G, however, a semicircular cover for the track tension adjusters was introduced. These coverings attached to the rear armor plate with a snap fit system.

Smoke Grenade Box Development &
Track Tension Cover

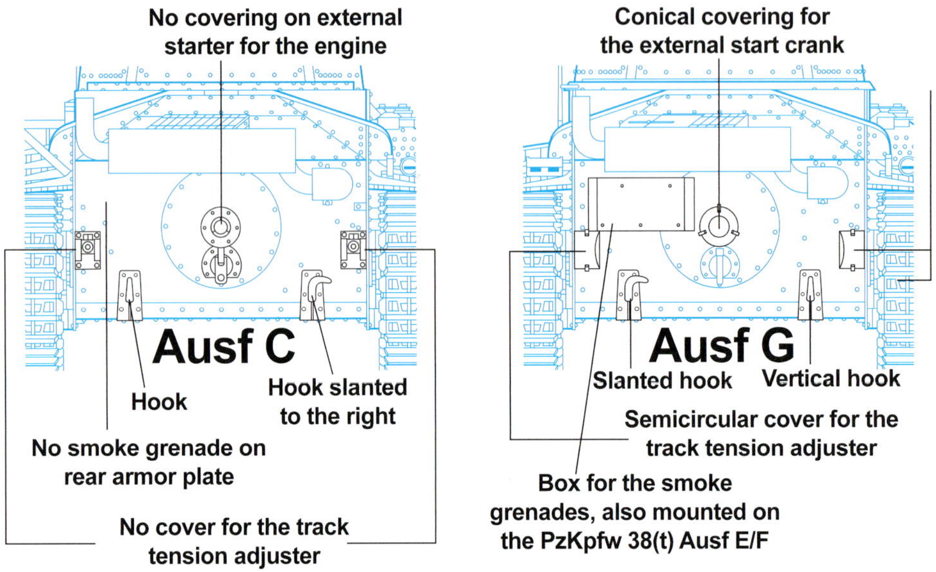

No covering on external starter for the engine

Ausf C

Hook

Hook slanted to the right

No smoke grenade on rear armor plate

No cover for the track tension adjuster

Conical covering for the external start crank

Ausf G

Slanted hook   Vertical hook

Semicircular cover for the track tension adjuster

Box for the smoke grenades, also mounted on the PzKpfw 38(t) Ausf E/F

Six bolts held the tow hook to the circular rear cover plate for the fan. The tow hook is missing its safety catch on this particular vehicle, something that is not standard for operational PzKpfw 38(t)s.

Located above the tow hook is the shaft for the hand crank. The Praga TNHPS/II engine could be manually started with the help of a crank in case the Bosch electrical starter motor failed. This kind of shaft design shown on this vehicle in the Deutsches Panzermuseum became standard for the PzKpfw 38(t) Ausf A through Ausf F variants. The PzKpfw 38(t) Ausf G featured a conical covering over the crank. This cover was attached with a snap-fit system on the circular rear armored access panel for the fan.

Removing the 15 bolts around the circumference of the circular plate allowed access to the fan. This kind of cover plate was introduced on the Czechoslovak pre-war LT vz 38 and the PzKpfw 38(t).

Six rivets hold the left towing hook.

This hook is on the vehicle's right rear.

This shaft is for the hand crank as viewed from the left.

The shaft for the hand crank is viewed from the right. A single nut is located on the base of the circular armor plate, which is to the right of the shaft. No nut is on the left side of the shaft base.

Three Swiss Panzerwagen 39s (Armored Vehicle 39) cross a river during exercises in June 1942. The first modern tank in the Swiss Army's inventory was armed with a single 24mm Panzerwagen-Kanone 38 gun and two oil-cooled 7.5mm Panzerwagen Maschinengewehre 38 machine guns. Both weapons were developed by the Eidgenössische Waffenfabrik (Swiss Federal Arms Factory). Vehicle M-7561 has the hull number 17 and belongs to the Panzerwagen Detachment 6. (Swiss Federal Archives)

This Swiss Panzerwagen 39 belonging to Detachment 2 has a lobster emblem applied as a unit insignia on the turret. The small chevron in front of the unit insignia denoted the vehicle within the detachment. A single chevron stands for the first tank in the detachment. These emblems were removed from the vehicles when the detachments were reconfigured into a Kompanie (company). The shovel is missing on the superstructure of this particular Panzerwagen 39. The vehicle has a foldable antenna mast that had been attached on the right of the turret. (Swiss Federal Archives)

This Swiss Panzerwagen 39 (M-7542/hull number 10) of the Panzerwagen Detachment 4 (4th armored vehicle detachment) is taking part in exercises on 4 October 1940 and has a turtle applied as a unit marking on the turret. All Panzerwagen 39s had license plates with four-digit numbers. On all 24 vehicles, the registration number started with 75, followed by another two individual numbers. The letter "M" stands for Militär (army). Military license plates were painted on the lower front armor plate as well the rear armor plate of the tanks. (Swiss Federal Archives)

Panzerwagen 39 (M-7553/hull number 15) of the Panzerwagen Detachment 5 has the yellow *CH* neutrality marking. The type of horn located below the searchlight is Swiss in origin and was applied on vehicles assembled in Switzerland from components delivered from the Českomoravská-Kolben-Daněk (ČKD). The searchlight is slightly offset to the left so as not to interfere with the driver's field of view. (Swiss Federal Archives)

The yellow *CH*, a neutrality marking introduced on 14 June 1940, was applied on the upper surfaces of the turret, just around the fairing for the ammunition rack for the 24mm Panzerwagen-Kanone 38 gun. (Swiss Federal Archives)

Panzerwagen 39 (M-7512/hull number 2) of Panzerwagen Detachment 1 is one of 12 vehicles ČKD delivered to the Swiss Army in 1939. On exercises in June 1941, protective material covered the Panzerwagen-Kanone 38 gun barrel nozzle and the barrels of the 7.5mm Panzerwagen Maschinengewehr machine guns. (Swiss Federal Archives)

Neutrality markings appear on the sides and rear armor plate of the Panzerwagen 39 turret. The 2nd detachment's lobster unit insignia was applied between the *C* and *H* on the turret's rear. There is a round stop light above the military license plate—a rarity on Panzerwagen 39s. (Swiss Federal Archives)

This Panzerwagen 39 (M-7561/hull number 17) of Panzerwagen Detachment 6 takes part in an exercise in June 1942. This vehicle had been assembled in Switzerland from components delivered by the Českomoravská-Kolben-Daněk (ČKD) in February 1939. This particular tank is powered by a Saurer CT-1D diesel engine with an output of 110 hp. The muzzles of the cannon and the two machine guns were covered, a common feature on Swiss Panzerwagen 39s during World War II. (Swiss Federal Archives 09444)

This Swiss Panzerwagen 39 takes part in an exercise on 22 May 1942. No neutrality markings were applied on this particular tank, which is unusual for Swiss armored vehicles of that time. The engine bay access panels and the rear part of the hull were covered with mud for camouflage. The upper cover plate, on which the turret was welded, could be opened up to 30mm, giving the commander a better field of view than through the visor in the copula, as shown by this particular vehicle. This Panzerwagen 39 lacks the protective coverings on the barrels of the gun and the two machine guns, indicating that firing trials were part of the exercises. (Swiss Federal Archives 10149)

This Panzerwagen 39 (M-7553/hull number 15) of Panzerwagen Detachment 5 carrles a unit marking on the turret in front of the neutrality markings. A 7.5mm Leichtes Maschinengewehr 25 (LMG 25), rather than the standard Panzerwagen Maschinengewehr 38, is installed in the left front armor plate. The barrel of the Panzerwagen Maschinengewehr 38, mounted in the turret, is covered. The left circular reflector on top of the fender is clear, while the right reflector is red. The horn was mounted below the searchlight on the Swiss Panzerwagen 39, while the German PzKpfw 38(t) had the horn located on the right fender, just below the side visor. All Panzerwagen 39s were equipped with a search light, but it did not come with the German PzKpfw 38(t)s. The Swiss Panzerwagen had a single large visor for the driver on the right front plate, while the German PzKpfw 38(t) had two visors: an additional one for the radio operator was on the left. The visors of the German PzKpfw 38(t) and the Swiss Panzerwagen 39 were of different design. With the factory designation LTL-H, the vehicle was unofficially nicknamed "Praga" in the Swiss Army during its first period of service. On 16 January 1940, however, the Swiss Department of Defense issued an order banning the use of this nickname. (Swiss Federal Archives)

A very distinctive feature of the Swiss Panzerwagen 39 is the conduit attached on the base of each armored engine bay door. A slide is attached on each side of the of the square-shaped air exhaust tunnel for the radiator. These two slides regulated the air flow. (Swiss Federal Archives)

This Panzerwagen 39 (M-7514/hull number 4) of Panzerwagen Detachment 1 drives past a small village in the Western part of Switzerland during exercises. Condor motorcycles manufactured in Switzerland escort the Panzerwagen 39. (Swiss Federal Archives)

This Panzerwagen 39 (M-7511/hull number 1) is driven onto a transport trailer towed by a Swiss-built Saurer M6 heavy truck (M-7515). Two red circular reflectors were mounted on the front fender. (Swiss Federal Archives)

The Panzerwagen 39s were transported on flatcars between the proving grounds and the arsenals. The antenna mast has here been folded for transport. The tank is towed by a Saurer M6 six-wheel truck. Most Saurer M6 all-wheel-drive trucks served with artillery units of the Swiss Army, towing heavy cannons. (Swiss Federal Archives)

The rear view mirror housing is on the right front fender of this Swiss Panzerwagen 39 (M-7562/hull number 18). A blackout light is at the end of the cable attached with a spring on the upper edge of the hull front. By turning the cable, the light could also attach atop of the front fender. The horn under the searchlight has a flush head. (Swiss Federal Archives)

The turret of this Panzerwagen 39 (M-7544/hull number 12) features the turtle insignia of Panzerwagen Detachment 4. An L-shaped stand on the left fender holds a rectangular rear-view mirror. A spare cable wraps around the horn, which is of the type used on the German PzKpfw 38(t). This vehicle bears the original three-tone factory camouflage of earth brown, tan, and olive green. (Swiss Federal Archives)

This Panzerwagen 39 (M-7563/hull number 19) belonged to the Panzerwagen Detachment 6, which became the Panzerwagen Kompanie 3 (3rd armored vehicle company) in 1940. (Swiss Federal Archives)

This Panzerwagen 39 carries a two-tone irregular camouflage pattern of olive green and dark green. It is believed that the second batch of 12 Panzerwagen 39s assembled in Switzerland were painted in this two-tone camouflage. (Swiss Federal Archives)

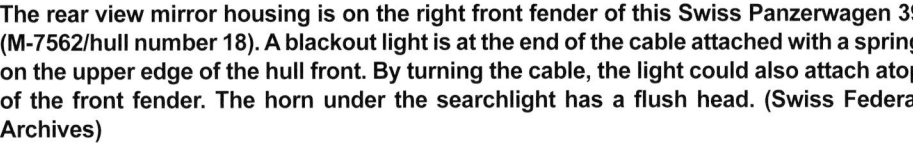

(Top Left) The Panzerwagen 39's suspension and tracks were both similar to what the early- production PzKpfw 38(t) had. A distinctive feature of the German PzKpfw 38(t) is the lack of a circular cover plate in the center of the road wheel, resulting in a circular recess in the center of the road wheel. The Swiss Panzerwagen 39 had three return rollers on each side. The rear one had a different face than the first two. All Swiss Panzerwagen 39s had a shovel and an axe stored on the right superstructure. The antenna socket was mounted on the right side armor of the turret. (Swiss Federal Archives)

(Top Right) A distinctive feature for the Swiss Panzerwagen 39 was the three return rollers attached on each side of the suspension. All German PzKpfw 38(t)s were equipped with only two return rollers, omitting the rear return roller. The Panzerwagen 39 had a hammer and a jack mounted on the left superstructure. In contrast to the German PzKpfw 38(t), the Panzerwagen lacked the square wooden jack base. This particular Panzerwagen 39 has a lobster painted as a unit emblem on both sides and the rear of the turret. The circular reflector attached on the conduit mounted on the armored engine bay door is a field modification. (Swiss Federal Archives)

(Bottom Right) This Panzerwagen 39 (M-7522/hull number 6) of Panzerwagen Detachment 2 crosses a trench during exercises. This particular tank has two white chevrons painted in front of the lobster, denoting that this was the second vehicle within the detachment. The chevron as an individual vehicle marking was only carried by Detachment 2. The 2nd detachment had initially chosen a scorpion as a unit emblem, but the unit artist did not know what a scorpion looked like, so a lobster was painted instead. (Swiss Federal Archives)

Three Detachment 2 vehicles move forward during an exercise. The first two tanks have chevrons as individual vehicle markings, which is a dedicated feature for Detachment 2. The lobster unit insignia is emblazoned on all three tanks. The Panzerwagen 39 has 20mm side armor plates on the turret and 10mm roof armor The commander's copula incorporates a visor. With the help of a ball bearing ring race, the copula and the visor could turn 360 degrees. The antenna masts on these vehicles are folded back. (Swiss Federal Archives)

A Panzerwagen 39 gets an overhaul in June 1941. The conduit attaches on the face of the armored engine-bay access doors. The conduit extensions lead into the drain pipe between the muffler and the rear fender. The Panzerwagen 39 has a shorter turret than the PzKpfw 38(t) and had to rotate its turret at a far lower angle than the 90 degrees the PzKpfw 38(t) had to rotate its turret to open the armored engine bay doors. Air for cooling the Saurer CT-1 engine was inducted through openings in each side of the compartment, protected by the overhang of the engine deck. (Swiss Federal Archives)

This vehicle (M-7563) passes through heavy mud during an exercise in June 1941. A circular red reflector is mounted unusually near the outer end of the fender brackets. The letter *C* of the neutrality marking on the rear armor plate of the turret ends in two bars. This application does not follow the guidelines for the application of the neutrality markings that were issued by the Supreme Command of the Swiss Army on 10 June 1940. The antenna mast is in a raised position. (Swiss Federal Archives)

This Panzerwagen 39 (M-7561) has the red circular reflectors mounted, as was standard practice, at the inner end of the fender brackets. Also standard is the application of the neutrality marking on the rear armor plate with the rounded ends of the letter *C*. The antenna mast is in a retracted position. The bag behind the cooling air exhaust for the radiator contains a canvas cover. (Swiss Federal Archives)

This Panzerwagen 39 (M-7511) has two red circular reflectors attached on the conduit during an exercise in June 1941. The position of the reflectors is a field modification. No circular reflectors were carried on the rear fender. This particular tank features the diamond-shaped double brake light housing that became standard on the Swiss Panzerwagen 39. A frame is bolted on the rear armor plate of the turret. M-7511 became the sole Panzerwagen 39 having the copula mounted on the left for trials. All other Panzerwagen 39s had the copula located on the right. The M-7511 was rebuilt with the standard copula configuration after the end of WWII. (Swiss Federal Archives)

This Panzerwagen 39 (M-7513/hull number 3), shown on exercises in June 1941, lacks the entire muffler and the metal cover cap on the right rear. The exhaust gases simply exit through the circular aperture. Non standard for the Panzerwagen 39 is the circular stop light, which was attached on only a few examples. The red reflector was attached on the conduit as a field modification. In contrast to the German PzKpfw 38(t), the Panzerwagen 39 lacks the large circular plate cut in the rear armor plate of the hull. The track tension adjusters on the Panzerwagen 39 were located in the inner hull while the German PzKpfw 38(t) had the track tension adjusters attached externally on both edges of the rear armor plate. (Swiss Federal Archives)

The housing for the rear-view mirror was mounted on the right front fender on the Swiss Panzerwagen 39, while all German PzKpfw 38(t)s had the rear-view mirror housing located on the left fender. The fender extension, made of leather, is a feature unique to the Swiss Panzerwagen 39 and the LTP for Peru; no PzKpfw 38(t) had any such extension.

A unique feature for the Swiss Panzerwagen 39 is the axe on the front 15mm armor plate. The electric cable for the blackout light was attached on the covering for the fender. This type of blackout light is of Czechoslovak origin and was applied on both fenders. Alternatively the blackout light could be also fixed on the front armor plate.

A blackout light was located on both front fenders of the Swiss Panzerwagen 39. The blackout light was attached to the fender with a spring. A number of early German PzKpfw 38(t)s were equipped with this type of blackout light, supplementing the standard German Notek blackout light. This type of circular white reflector was also mounted on the German PzKpfw 38(t).

The blackout light was attached with a spring to both front fenders on the Swiss Panzerwagen 39. When the blackout light was attached to the front armor plate, the same type of spring was used. The same type of blackout light and spring attachment was mounted on the LTP for Peru.

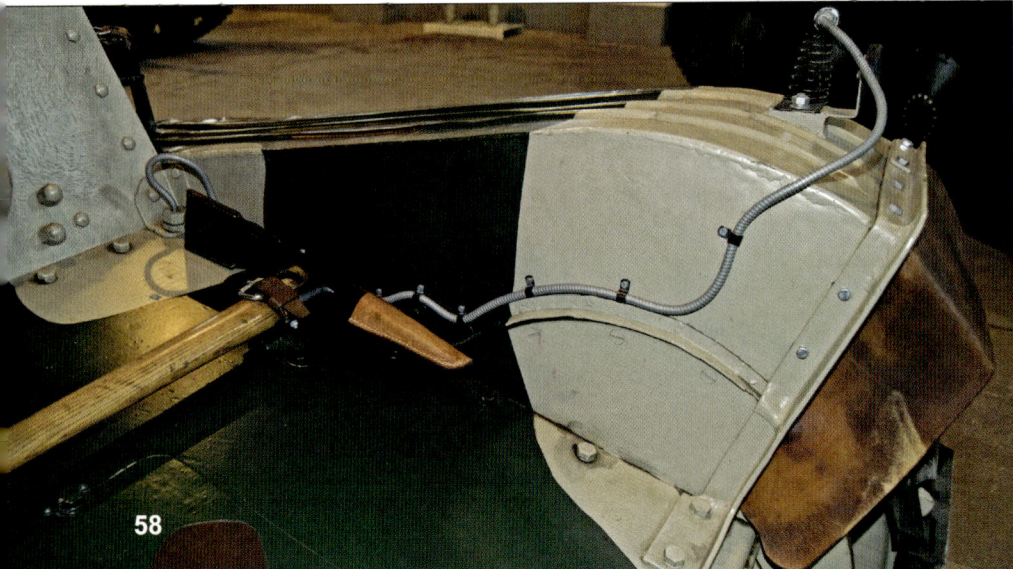

The oil-cooled 7.5mm Panzerwagen Maschinengewehr 38 weighed 18 kilograms and had a rate of fire of 1,200 rounds per minute. A unique feature of the Swiss Panzerwagen 39 was the search light, underneath which the horn was mounted. This type of horn with a flush head is a typical feature for Panzerwagen 39s assembled in Switzerland. All Panzerwagen 39s built in Czechoslovakia had a ring on the head of the horn.

The Panzerwagen Maschinengewehr 38 was built in a ball mount. This oil-cooled weapon was developed by the Eidgenössische Waffenfabrik (Swiss Federal Arms Factory) at Bern and only saw service with the Panzerwagen 39. This cooling system required a coolant jacket. The machine gun was 990mm long. The nearly vertical armor plate protecting the driver had a thickness of 25mm.

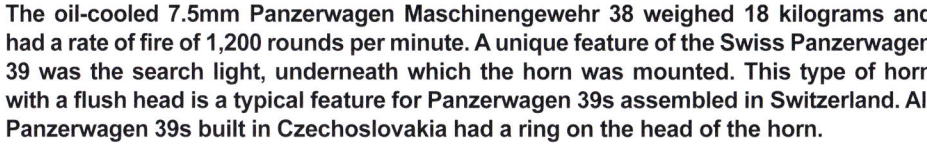

In contrast to the German PzKpfw 38(t), the Swiss Panzerwagen 39 only had a single visor in the front plate, which was mounted on the right side. The visor of the Panzerwagen 39 had side slits not present on the PzKpfw 38(t). Blocks of armor glass could be inserted to protect the driver. The same type of visor was built into the LTP for Peru.

These two handles closed an armor plate, which protected the armor glass blocks that were mounted in the visors. These armor plates gave the driver additional protection. Not visible are two additional handles that fix the visors to the front armor plate when fully closed and prevent the visors from being opened due to enemy small arms fire.

The Swiss Panzerwagen 39 and the German PzKpfw 38(t) both used the same type of road wheel with rubber tires. A typical feature for the Panzerwagen 39 was the introduction of a third return roller and the lack of a circular cover plate in the center of the road wheel. For this reason, the road wheels of the Panzerwagen 39 had a recess in the center of the wheel. The LTP tank for the Peruvian Army used the same type of road wheel as well a third return roller.

The center of the road wheel of a Panzerwagen 39 with the circular recess was typical for the Swiss LTL-H and the Peruvian LTP. The nipple in the center of the road wheel was used to lubricate the bearings with grease. Initially the Swiss Army suffered difficulties in properly greasing the bearings of the road wheels since the grease pistols imported from France in early 1939 lacked the necessary pressure for proper lubrication of the wheels. As a result, the Swiss Army developed its own grease pistols.

The sprocket wheel of a Swiss Panzerwagen 39 had 12 circular apertures in the outer part of the sprocket while the inner part of the sprocket was solid. This same type of sprocket was mounted on the LT-40, built during 1940 for the Slovak Army, and the LTP, built for Peru. The PzKpfw 38(t) only had eight of these circular apertures and featured eight arch-shaped apertures in the inner part of the sprocket.

The two return rollers at the front and in the middle featured a metal ring on their heads. The wheels of these two return rollers of the Panzerwagen 39 were of slightly smaller diameter than was the third return roller mounted on the rear.

The rear idler wheel of a Panzerwagen 39 had circular apertures in the outer section. Idler wheels with circular apertures were never used on the PzKpfw 38(t), whose wheels featured apertures shaped like figure eights. Eight small circular apertures were located in the inner idler wheel of the Panzerwagen 39. German PzKpfw 38(t) idler wheels lacked these eight inner apertures and instead had a thicker circular cover plate. An idler wheel identical to that on the Panzerwagen 39 was used on the LTP for Peru and the LT-40 for the Slovak Army.

There was a rear return roller attached to the suspension on both sides of the Panzerwagen 39. Unlike the front and middle return wheels of the suspension, the rear roller had no metal ring around its center. The diameter of the rubber-tired wheel was slightly larger. The two return rollers on the German PzKpfw 38(t) were nearly identical to the rear return roller on the Panzerwagen 39, except that their rubber tires were thicker.

The track tension adjuster on the Swiss Panzerwagen 39 was located inside the rear hull with an adjustment bolt protruding on the base of the rear armor plate. This bolt was located on each side between the tow hook and the idler wheel. On all German PzKpfw 38(t)s, the two track tension adjusters were attached externally on the upper rear armor plate.

A shovel and an axe are mounted on the right-side superstructure. This tool arrangement was not used by any other tank type built by ČKD. Leather straps secure these tools. Below the axe handle is the driver's right-side visor. The German PzKpfw 38(t) had a housing mounted on the visor, not only a horizontal slit as on the Panzerwagen 39.

A jack and hammer are mounted to the left-side superstructure. These tools were located just below the turret. The jack and hammer were both secured with leather straps.

The shovel was secured in a crescent-shaped support that was riveted to the superstructure of the Swiss Panzerwagen 39.

In contrast to the German PzKpfw 38(t), no Swiss Panzerwagen 39 was equipped with the square-shaped wooden jack base. The wooden jack base, a basic piece of equipment on the PzKpfw 38(t), insured that the jack would not sink into soft ground during repair or replacement of a road wheel.

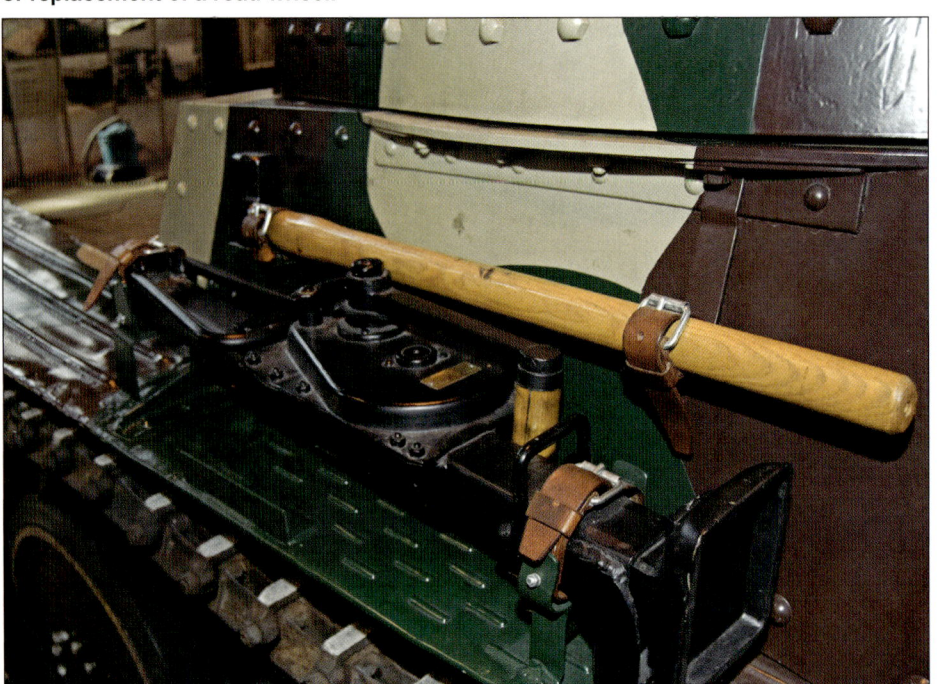

The Swiss-designed 24mm Panzerwagen-Kanone 38 was developed especially for the Panzerwagen 39 vehicle by the government-owned enterprise Eidgenössische Waffenfabrik (Swiss Federal Arms Factory) at Bern. The Panzerwagen 39 carried an ammunition supply of 114 rounds for the Panzerwagen-Kanone 38 cannon as well a total of 1,200 rounds for the two 7.5mm Panzerwagen Maschinengewehr 38 machine guns.

A sting attached on the barrel of the Panzerwagen-Kanone 38 prevented damage from the rounds fired by the Panzerwagen Maschinengewehr 38. The gun and the single machine gun in the turret were connected by a linkage system. The machine gun mounted in the superstructure of the vehicle did not have a linkage system. The commander could, however, operate the machine guns in the turret independently.

The muzzle belongs to the 1,880mm long barrel of the Panzerwagen-Kanone 38, a weapon with a 24mm caliber. This type of muzzle with its fairing of the muzzle brake is unique to the Swiss Panzerwagen 39 since the KwK 38(t) 37mm cannon of the PzKpfw 38(t) had a flush muzzle without a muzzle brake. The Panzerwagen-Kanone 38 had a muzzle velocity of 900 meters per second that enabled this efficient weapon to penetrate armor plates up to 30mm thick at a range of 200 meters.

A distinctive feature for the Panzerwagen 39 and the Peruvian LTP was the overhang of the upper rear turret decking. The German PzKpfw 38(t) lacked such an overhang, but the tank had an extended turret that allowed the stowage of additional ammunition in the rear of the turret.

The antenna platform was attached on the right side of the turret. In contrast to the fixed antenna mast that was mounted on the left front superstructure of the German PzKpfw 38(t), the antenna mast of the Panzerwagen 39 was foldable.

Built into the access hatch on the right side of the turret was a small observation copula for the commander. Attached to the copula is a fixture for installing a Leichtes Maschinengewehr, a small-caliber machine gun intended for anti-aircraft defense purposes.

A two-piece entrance door for the crew was located on top of the turret. In contrast to the PzKpfw 38(t), the Swiss Panzerwagen 38(t) lacked the second entrance hatch located on top of the left superstructure over the radio operator position. A leather cushion is on the inner surface of the door above the gunner.

When the Panzerwagen-Kanone 38 was at maximum depression, the magazine interfered with the ceiling plate of the turret. A spring-loaded opening panel was introduced in the turret's roof to compensate and also to provide ventilation during firing.

The copula for the commander incorporated a ball bearing race so that the commander could turn the single visor 360 degrees. The German PzKpfw 38(t) instead had four fixed visors as well a coaxial periscope in the commander's copula.

A visor is mounted on both sides of the Panzerwagen 39's turret. To protect the crew, these visors had bulletproof armor glass blocks that could be replaced after suffering damage. The lever fit into an armor plate that protected the armor glass block

The Swiss Panzerwagen 39 had a crew of three. The position of the gunner, who also operated the radios, was on the left side in the turret. The German PzKpfw 38(t) had a crew of four with the radio operator located beside the driver. In the Swiss Panzerwagen 39 the commander was located on the right while the commander's position was on the left in the German PzKpfw 38(t). The locking mechanism for the main turret entrance door is equipped with a shock absorber. The right side visor is located in the inner armor plate of the turret. The side armor plates of the turret had a thickness of 20mm, while the front armor was 30mm thick. The Swiss turret had thicker armor than the early models of the PzKpfw 38(t) turret, with their 15mm side armor and 25mm front armor.

A Swiss cross, the manufacturer's name, and a serial number are stamped on the breech of the Panzerwagen-Kanone 38 that is mounted in the Panzerwagen 39 when viewed from above. Českomoravská-Kolben-Daněk delivered all 24 Panzerwagen 39s without any armament.

The breech of the Panzerwagen-Kanone 38 is viewed from the right. The spent cases were collected in a leather bag mounted below the weapon, which weighed 77 kilograms and was 2,590mm long. The rate of fire was 30 to 40 rounds per minute.

This view shows the left inner turret position for the gunner with the seat and the hand wheel for setting the elevation of the Panzerwagen-Kanone 38.

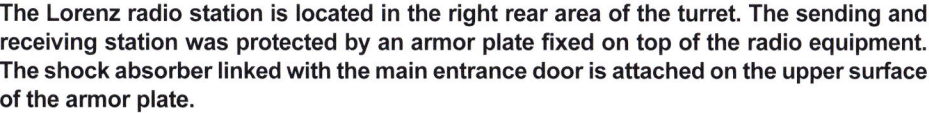

The gunner's station is on the left side, inside the turret. A spare magazine and visor are attached to the armor plating.

The ammunition magazine for the Panzerwagen-Kanone 38 gun is attached to the right inner armor plate. Each rack contained 6 rounds.

The Lorenz radio station is located in the right rear area of the turret. The sending and receiving station was protected by an armor plate fixed on top of the radio equipment. The shock absorber linked with the main entrance door is attached on the upper surface of the armor plate.

A first aid kit is strapped to the inside of the armor plate by the Lorenz radio station in the right rear of the turret. The radio sets were fitted into the vehicles in Switzerland. Českomoravská-Kolben-Daněk delivered the LTL-H without radio equipment or armament. The Lorenz radio station had a range of about 10 kilometers.

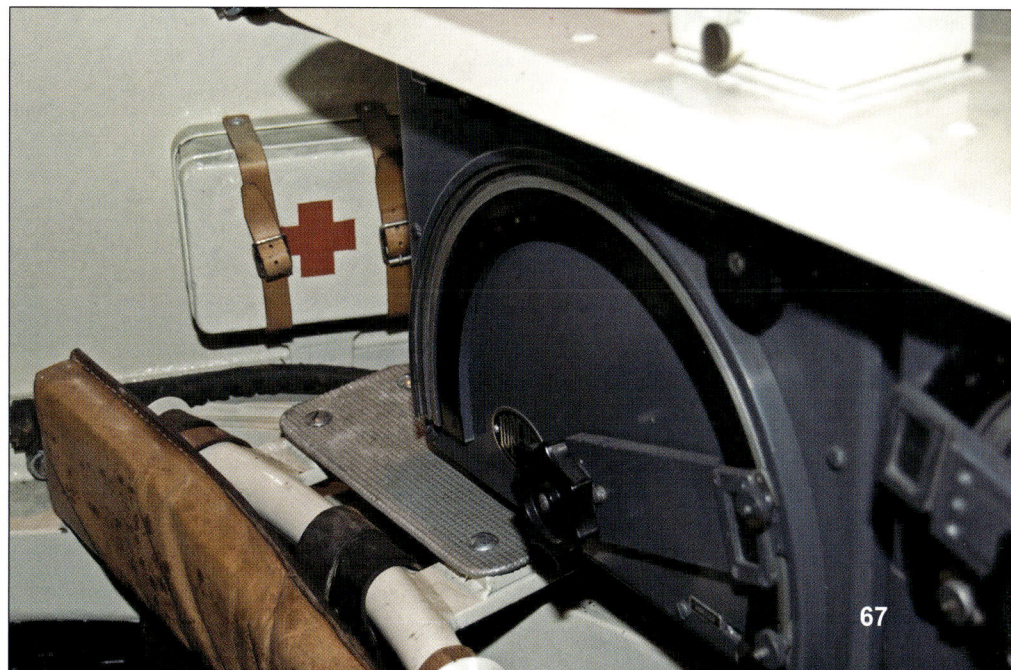

The left engine access panel is open. These doors had 15mm-thick armor protection. The engine access panels on the German PzKpfw 38(t) were similar but only had 8mm of armor protection. The Swiss panels had three hinges to connect these doors to the engine decking while the German panels only had two hinges.

At the right rear inside the engine compartment are the drum-shaped air filter and the radiator. The layout for these components was similar on the German PzKpfw 38(t).

Two silver drums located to the right in the Panzerwagen 39 engine compartment are the air filters for the Saurer CT-1D diesel engine. The engine arrangement of the German PzKpfw 38(t) was nearly identical to that of the Swiss Panzerwagen 39.

Visible in the right front of the engine compartment is one of two hatches with adjustable vents that were built into the firewall separating the crew compartment from the engine compartment.

The Saurer CT-1D diesel engine, seen here from the left, was a derivative from the Scania-Vabis 1664 truck engine. This diesel powerplant was developed by Českomoravská-Kolben-Daněk in cooperation with the Swiss engine manufacturer Aktiengesellschaft Adolph Saurer. The diesel engine offered 110 hp at 2,200 rpm. A prototype engine was tested in Switzerland during June 1938.

The diesel engine, seen here in the left rear engine compartment of a Panzerwagen 39, proved to be more powerful than the gasoline powerplant at low rpm levels. The Swiss Army became the sole operator of diesel engines in the vehicle. The Saurer CT-1D diesel engine could start electrically or manually, with the help of a fly wheel.

Two hatches with adjustable vents are located at the front left of the engine compartment and allowed the crew access to the engine compartment.

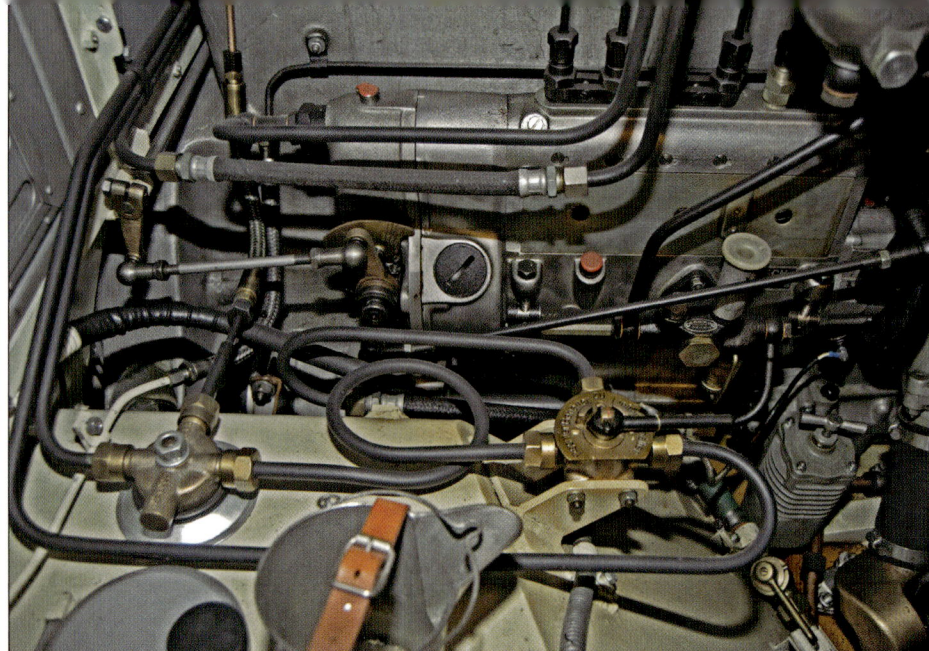

The left 110-liter fuel tank located beside the engine block appears in the front lower-left engine compartment. The wiring and the arrangement of the tubes in the engine compartment were different on the diesel-powered Panzerwagen 39, but the tanks for the diesel- and gasoline-powered Panzerwagen 39s were identical, apart from the fact that the diesel tanks were pressurized.

The Panzerwagen 39 has a 110-liter fuel tank, shown here. An identical tank was on the right side. This tank configuration was nearly identical on the German PzKpfw 38(t).

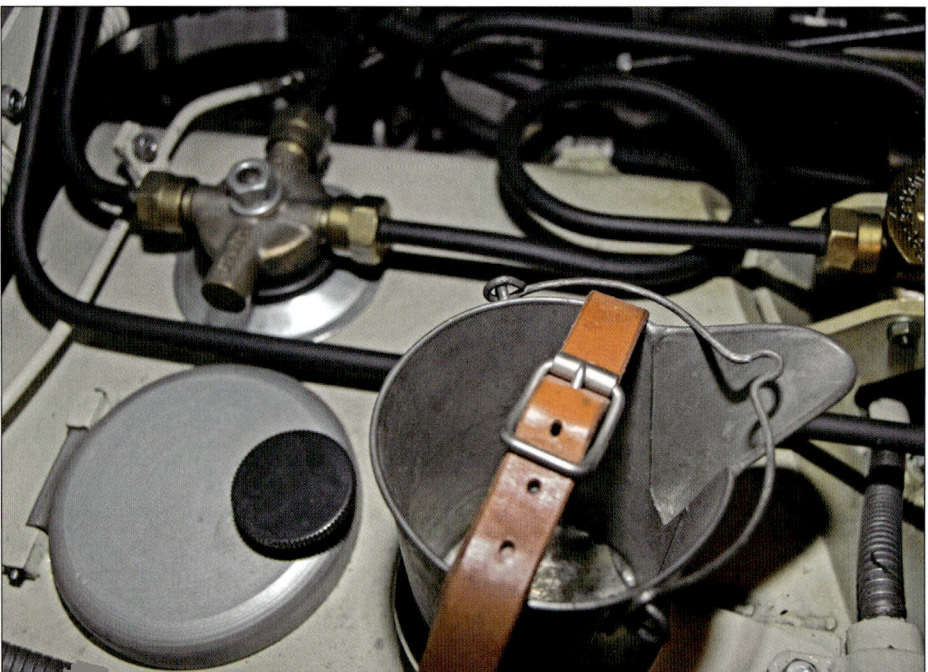

These engine accessory components belong to the Saurer CT-1D diesel engine. The Saurer CT-1D diesel engine was originally intended to equip all Panzerwagen 39s, but since the diesel powerplant was not ready for production on time, the Swiss Army equipped the first 12 vehicles with the Saurer CT-1 gasoline engine.

This left engine cylinder block is for the Saurer CT-1D diesel engine. The BMM factory built this engine, designated Praga TNHPS/II, for the PzKpfw 38(t). The engine was a four-stroke, six-cylinder, inline engine with a swept volume of 7.75 liters and a compression ratio of 1:6.2.

The pump for the coolant water is in the left engine compartment, just behind the fuel tank.

This view reveals the lower front of the engine and part of the left fuel tank. To properly distinguish gas- and diesel-powered vehicles, the Swiss Army issued an order that the suffix *mit Benzinmotor* (with gasoline engine) or *mit Dieselmotor* (with diesel engine) should be added behind the designation Panzerwagen 39 (armored vehicle 39).

The lower part of the pump for the coolant water is seen here. A shaft connects the pump to the engine.

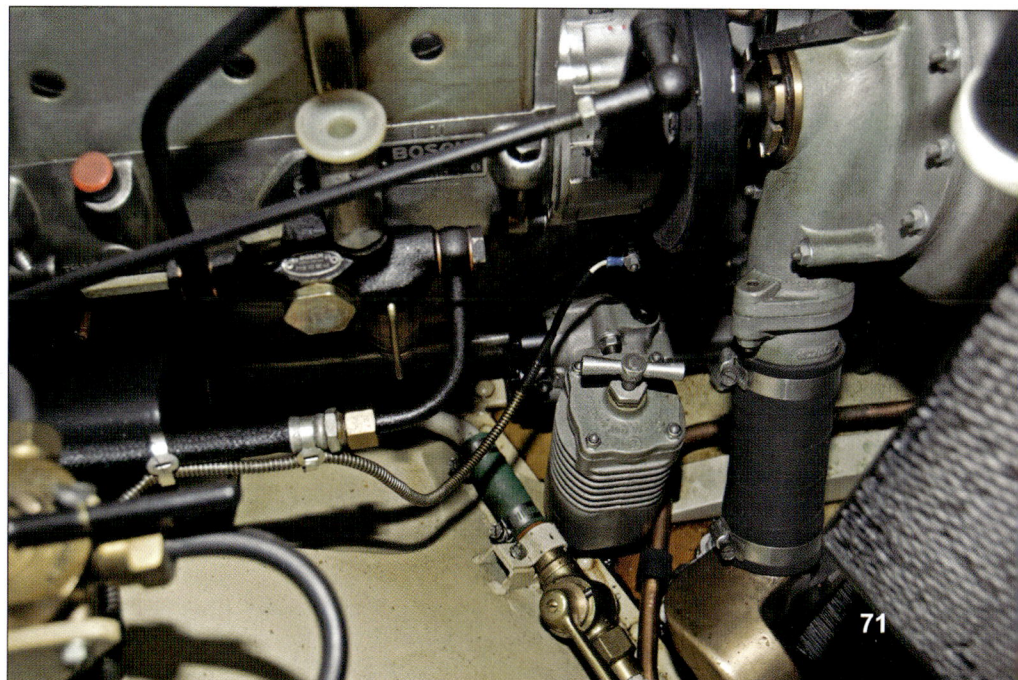

This PzKpfw 38(t) Ausf A, "White 223," saw action during the Polish campaign with the 3. Leichte Division (3rd Light Division), 67. Panzer Abteilung (67th Armored Battalion). The vehicle is camouflaged in Dunkelgrau RAL 7021 (dark grey) overall. The national marking and the tactical number were applied in Cremeweiss RAL 9001 (white).

This tank is an early PzKpfw 38(t) Ausf A that belonged to the 25. Panzer Regiment (25th Armored Regiment), 7. Panzer Division (7th Armored Division) and is painted in Dunkelgrau RAL 7021 (dark grey), which became standard for German tanks in the early years of the war. The large tactical numbers are a feature for the 7. Panzer Division.

This PzKpfw 38(t) Ausf B belonged to the 8. Panzer Division and took part in the invasion of the Soviet Union in summer 1941. This tank is equipped with three bins. Sand and mud had covered the original camouflage of Dunkelgrau RAL 7021 (dark grey). The tactical markings on the turret were applied in Maisgelb RAL 1006 (yellow). The "Y" with a bar is the divisional insignia for the 8th Armored Division and was introduced in 1941. The Roman number three stands for the 3. Abteilung (3rd Battalion). These markings were typical during the invasion of the Soviet Union.

This PzKpfw 38(t) Ausf C (Befehlswagen 38) carries the conspicuous frame aerial for long range communications, indicating a command vehicle. The marking R02 is applied in white. The letter "R" was assigned to regimental staff vehicles. The R02 denotes the regiment assistant to the commander. This Befehlswagen 38 belonged to the 27. Panzer Regiment, 19. Panzer Division. The vehicle is painted in standard Dunkelgrau RAL 7021 (dark grey) overall.

This PzKpfw 38(t) Ausf C, "Red 524" saw service on the Eastern front with the 7. Panzer Division. This vehicle is painted in Dunkelgrau RAL 7021 (dark grey). The 500 series on the vehicle numbers was assigned for regimental or battalion staff vehicles.

This PzKpfw 38(t) Ausf C still served in December 1944 on the Western front near the Saar river. The veteran vehicle received a frontline applied camouflage of Dunkelgelb RAL 7028 (dark yellow) and Olivgrün RAL 6003 (olive green) sprayed over the factory-applied Dunkelgrau RAL 7021 (dark grey).

A very distinctive feature on the Swiss Panzerwagen 39 was the conduit attached to the base of the left and right armored engine bay doors. The drain pipe is located on the rear end of the conduit, between the covering for the muffler and rear fender.

In order to prevent water or burning liquids from Molotov cocktails from getting into the engine, drainage conduits were bolted on both faces of the engine bay doors and reinforced with bolted brackets.

The ČKD factory equipped all Panzerwagen 39s with conduits on the base of the engine bay doors according to Swiss Army specifications. These conduits prevented water from leaking in from the overhang of the engine cover doors that were designed to let in air. All Panzerwagen 39s had two spare track links attached on top of the left rear fender.

On the centerline of the engine decking was a watertight bag that held the vehicle's canvas cover. The canvas cover protected the vehicle from rain and snow in the open field.

All Panzerwagen 39s had an X-shaped rib stamped in the rear fender, a contrast to the four longitudinal ribs in the fenders of the German PzKpfw 38(t). The feature was intended to enhance the stability of the fender. A circular red tail reflector is attached on the bracket. Such red reflectors were also carried by the German PzKpfw 38(t).

This box, wrapped in a watertight fabric covering, contained special tools for the proper setting of the track tension. All Panzerwagen 39s carried this tool box on the right fender near the rear of the vehicle. The track tension adjusters were located on the lower rear armor plate of the Panzerwagen 39.

The X-shaped reinforcing stamp is visible on the left rear fender. The circular reflector could be detached from the L-shaped bracket by removing the nut from the bolt in the reflector. On some of the Swiss Panzerwagen 39s, the circular reflector was fixed in the inner position on the bracket rather than the outer position, as seen here.

Two spare track links were bolted in front of the L-bracket that connected the fender with the superstructure. The rear fender has two embossed Xs. A drain pipe can be seen at the rear end of the conduit. The conduits on the left and right sides of the vehicle had separate drain pipes.

This PzKpfw 38(t) Ausf E carries a single digit, yellow tactical number. Single digit tactical numbers were unusual among Panzer Divisions of the Wehrmacht. This particular vehicle belonged to the 20. Panzer Division (20th Armored Division) and saw action in the Russian campaign. The vehicle is painted in Dunkelgrau RAL 7021 (dark grey) overall. The tactical number is applied in Maisgelb RAL 1006 (yellow). Single digit numbers are a feature for the 20th Armored Division.

This PzKpfw 38(t) Ausf E belonged to the 204. Panzer Regiment (204th Panzer Regiment), 22. Panzer Division (22nd Armored Division). This vehicle was delivered in the standard camouflage of Dunkelgrau RAL 7021 (dark grey), but it received a new camouflage pattern on the front that was Dunkelgelb RAL 7028 (dark yellow).

This PzKpfw 38(t) Ausf E has a large, three digit tactical number. The vehicle is painted in Dunkelgrau 7021 (dark gray) overall. The vehicle number is applied in Feuerrot RAL 3000 (red) and a large, white outline. The PzKpfw 38(t) Ausf E saw action with the 7. Panzer Division (7th Armored Division) in France, where it was captured. This vehicle has a non-standard stowage bin attached on the rear turret. A handle was welded on each cover door of the engine as a field modification.

This PzKpfw 38(t) Ausf G, "White 522," belongs to the 22. Panzer Division (22nd Armored Division) and saw action in the campaign against the Soviet Union. The vehicle is painted in standard Dunkelgrau RAL 7021 (dark grey). Totally not standard for late production PzKpfw 38(t)s is the rear view mirror on the port, front fender.

This PzKpfw 38(t) Ausf G, "Red 1003," belonged to the 7. Panzer Division that took part against the Soviet Union. Additional stowage bins are a typical feature for vehicles assigned to the Eastern Front. This vehicle is painted in Dunkelgrau RAL 7021 (dark grey) overall. The vehicle number is painted in Feuerrot RAL 3000 (red) and outlined with Cremeweiss RAL 9001 (white). The four-digit number on the turret denotes a headquarters tank company.

This PzKpfw 38(t) Ausf G belonged to the 21. Panzer Regiment (21st Armored Regiment), 20. Panzer Division (20th Armored Division) and saw action in the Demyansk pocket. The inscription reads, "Oblt. v. Jagow" and probably refers to a killed officer, 1st Lieutenant von Jagow. Patches of Dunkelgelb RAL 7028 (dark yellow) had been applied on the Eastern front over the factory-applied Dunkelgrau RAL 7021 (dark grey).

Two slides were attached beside the square air exhaust on the rear of the engine decking of the Panzerwagen 39. These two slides are a unique feature on the Panzerwagen 39. Unlike the square air exhaust on all Swiss Panzerwagen, the air exhaust on the PzKpfw 38(t) was rhomboidal, and the only German model of the vehicle to feature similar slides was the PzKpfw 38(t) Ausf G.

This curved exhaust stub extension was mounted during 1988/89, when this particular vehicle was restored by members of the Armeefahrzeug Park (army vehicle depot) at Thun. The extension had been fixed with a muffle on the original straight stub. The exact purpose for this modification on this ready-to-run vehicle is not known.

Swiss Panzerwagen 39s were delivered with a short, straight horizontal exhaust stub. This curved extension was attached on the exhaust stub when this particular vehicle was restored between 1988 and 1989 and is not authentic. A curved exhaust stub was standard on all German PzKpfw 38(t)s from the beginning, but in contrast to this Swiss Panzerwagen 39, the exhaust stub was turned upward on all PzKpfw 38(t)s.

The metal covering for the muffler that was located on the right rear armor plate of the Panzerwagen 39 is here seen from below. An identical type of covering was used on the German PzKpfw 38(t). A drain pipe, a distinctive feature on the Swiss Panzerwagen 39, runs between the covering and the rear fender.

German armored vehicles lacked Wehrmacht license plates, but all Swiss Panzerwagen 39s received a military license plate. This particular example on exhibit in the hangar of the Collection of Historical Armor in the Swiss Army at Thun has the registration M-7573. Swiss armored vehicles were equipped with a stop light set in a rounded diamond-shaped housing that was added after the arrival of the tanks in Switzerland. Unlike the German Wehrmacht, which equipped its vehicles with the Notek convoy light, the Swiss Army did not use convoy lights on its armored vehicles.

A rounded diamond-shaped housing for the stop light was mounted on most Swiss Panzerwagen 39s. It was equipped with two lights: a stop light as well a rear light for night drive. A few Panzerwagen 39s were equipped with a circular stop light taken from Swiss Army trucks. The housing, as well the circular stop light, were of Swiss origin. Below the housing is the black license plate. The top corners of all wartime license plates were cropped. In addition to the individual white registration number, the plate also featured the Swiss coat of arms and the red letter "M" for Militär (army).

The towing hock of a Panzerwagen 39 was attached directly on the rear armor plate of the vehicle. The towing hock on the PzKpfw 38(t) was set at a higher position on the vehicle and welded to the circular armor cover plate for the fan. The tube above the towing hock is the external starter unit for the Saurer CT-1D diesel engine. The engine could be started by turning a hand crank activating a fly wheel.

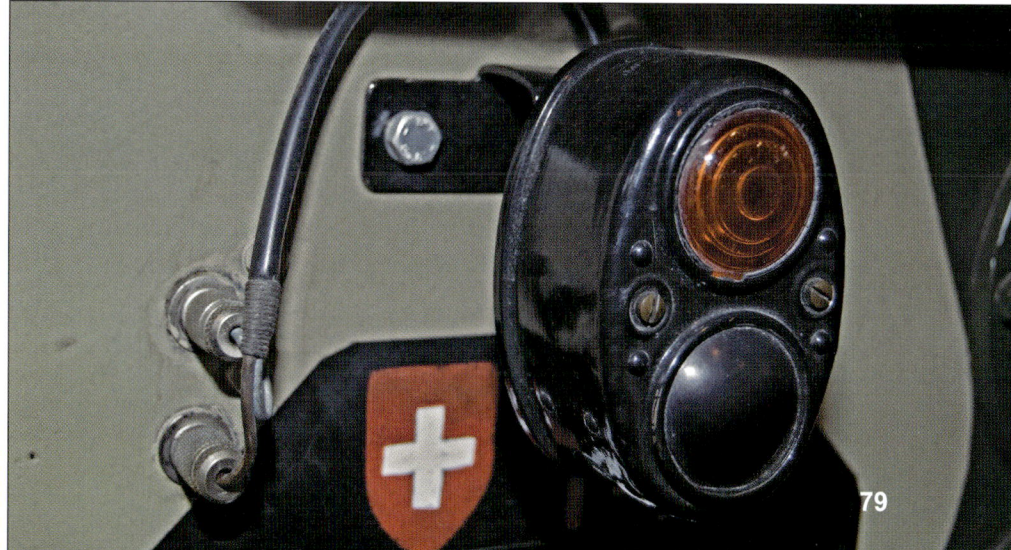

This Panzerwagen 39 (M-7562/hull number 18) is on exhibit at the Panzermuseum (Armor Museum) at Thun, Switzerland, and carries the highly visible neutrality markings that were introduced on 14 June 1940. The rhinoceros unit insignia on the turret indicates that the vehicle once belonged to the Panzerwagen Detachment 6. The lettering *CH*, Switzerland's international vehicle registration code, was applied in yellow to the hull front, and also to the roof, both sides, and rear armor plate of the turret. This marking was chosen after the Wehrmacht had applied large white crosses on its tanks as German national markings. A white cross on a red background, which was the national insignia that the Swiss emblazoned on Fliegertruppe (Swiss Air Force) aircraft, would have looked too similar to the German national markings on tanks. For this reason the yellow *CH* lettering was chosen. For a short period the Supreme Command of the Swiss Army had considered white lettering, but dropped this idea in favor of yellow. Switzerland's Supreme Command expected a German Wehrmacht invasion following the fall of France during the summer of 1940. A document signed by Henri Guisan, the Commander-in-Chief of the Swiss Army, states that "markings on Swiss tanks must be clearly distinguishable from enemy vehicles or vehicles belonging to the Allies." Although neutral throughout all of WWII, Switzerland considered the Third Reich to be a hostile nation during the first half of that conflict.